Preparing and Training for the Full Spectrum of Military Challenges

Insights from the Experiences of China, France, the United Kingdom, India, and Israel

David E. Johnson • Jennifer D. P. Moroney • Roger Cliff
M. Wade Markel • Laurence Smallman • Michael Spirtas

Prepared for the Office of the Secretary of Defense

Approved for public release; distribution unlimited

NATIONAL DEFENSE RESEARCH INSTITUTE

The research described in this report was prepared for the Office of the Secretary of Defense (OSD). The research was conducted in the RAND National Defense Research Institute, a federally funded research and development center sponsored by the OSD, the Joint Staff, the Unified Combatant Commands, the Department of the Navy, the Marine Corps, the defense agencies, and the defense Intelligence Community under Contract W74V8H-06-C-0002.

Library of Congress Cataloging-in-Publication Data

Preparing and training for the full spectrum of military challenges : insights from
    the experiences of China, the United Kingdom, France, India, and Israel /
    David E. Johnson ... [et al.].
        p. cm.
    Includes bibliographical references.
    ISBN 978-0-8330-4781-6 (pbk. : alk. paper)
    1. Military planning. 2. Operational readiness (Military science) 3. Soldiers—
Training of. I. Johnson, David E.

U150.P74 2009
355.6'84—dc22

                                                                                    2009049933

The RAND Corporation is a nonprofit research organization providing objective analysis and effective solutions that address the challenges facing the public and private sectors around the world. RAND's publications do not necessarily reflect the opinions of its research clients and sponsors.
RAND® is a registered trademark.

Flags on cover courtesy of the Central Intelligence Agency.

Published 2009 by the RAND Corporation
1776 Main Street, P.O. Box 2138, Santa Monica, CA 90407-2138
1200 South Hayes Street, Arlington, VA 22202-5050
4570 Fifth Avenue, Suite 600, Pittsburgh, PA 15213-2665
RAND URL: http://www.rand.org/
To order RAND documents or to obtain additional information, contact
Distribution Services: Telephone: (310) 451-7002;
Fax: (310) 451-6915; Email: order@rand.org

Preface

The RAND Corporation was asked to analyze how China, France, the UK, India, and Israel approach training for full-spectrum operations and deployments. This monograph should be of interest to those concerned with military training requirements. Material in the text was current as of October 2008, when research for the study was completed.

This research was sponsored by the Office of the Secretary of Defense (OSD) and conducted within the Forces and Resources Policy Center of the RAND National Defense Research Institute, a federally funded research and development center sponsored by the Office of the Secretary of Defense, the Joint Staff, the Unified Combatant Commands, the Navy, the Marine Corps, the defense agencies, and the defense Intelligence Community.

For more information on RAND's Forces and Resources Policy Center, contact the Director, James Hosek. He can be reached by email at james_hosek@rand.org; by phone at 310-393-0411, extension 7183; or by mail at the RAND Corporation, 1776 Main Street, P.O. Box 2138, Santa Monica, California 90407-2138. More information about RAND is available at www.rand.org.

Contents

Figures

Tables

Summary

> I'm worried that we're losing the edge on our ability to conduct
> full-spectrum operations and major combat operations. . . . Some
> people say that we're so busy all we can do is focus on COIN
> [counterinsurgency] operations and we have no time to focus on
> major combat operations and I think that is wrong.
> —*Lieutenant General Rick Lynch, Commanding General,*
> *III Armored Corps*[1]

The difficult and continually evolving operations in Iraq and Afghanistan show the complexities of what is now termed *irregular warfare* and
highlight the need for new approaches to the security challenges with
which the United States is now contending and will likely confront
in the future. The research reported in this monograph focused on
answering a rather straightforward, but thus far largely unanswered,
question: What can the U.S. military learn from other militaries about
how better to prepare for full-spectrum operations and deployments?
To this end, RAND was asked by the OSD for Personnel and Readiness to examine the militaries of China, France, the UK, India, and
Israel.

Not surprisingly, the training and organizing approaches of the
armed forces of China, France, the UK, India, and Israel reflect the
demands placed on them by their specific strategic environments. Our
research, therefore, focused on identifying areas in which these coun-

---

[1] Kate Brannen, "Ft. Hood Commander Concerned Army Is Losing Full-Spectrum Capabilities," InsideDefense.com, April 9, 2009.

tries employ different approaches to readiness and operational issues that may offer potential benefits to the U.S. system.

Most Insights Come from Ground-Force Experiences

Early in the course of our research, it became obvious that the differences between how the United States and other nations train their ground forces are much greater than the differences between how they train their air and naval forces. The ways in which air and naval operations are conducted are much less affected by changes in the geographic and sociopolitical setting than are ground-force operations. Therefore, there is less scope for differences in how countries approach the problem of how to prepare air and naval forces for different contingencies. The principal adaptations required of air and naval forces are those dictated by (1) the relative capabilities of an adversary and (2) the specific rules of engagement imposed by the national command authorities. The adaptations of air or naval forces required by changes in the physical environment and the sociopolitical milieu are much less demanding than those required of ground forces. This is not to say that the former set of adaptations is not demanding: We only wish to note that U.S. air and naval forces face challenges comparable to those faced by the other nations and that all those forces train and prepare for the challenges in similar ways. The major difference is that the U.S. naval and air forces are much larger and their training is generally better resourced. The principal area of commonality shared by all the air forces we examined is the difficulty of integrating those air forces with ground forces. This is partly an issue of interservice cooperation and different perspectives, but it is also one of meaningful joint training in peacetime.

Ground forces face a very different situation than do naval and air forces. All of the states we examined, with the exception of China, are or have recently been engaged in active military operations that range from participation in large-scale combat operations to COIN to peacekeeping to train, advise, and assist (TAA) missions. These very different types of operations, in our view, suggest that ground forces face the greatest demands in preparing for multiple types of military challenges. Furthermore, several of the nations we examined take different approaches than does the United States and therefore yield the majority of the insights in this monograph.

Strategic Imperatives, the Range of Military Operations, Specialty Forces, and Human Capital

Each of the states we assessed organizes and maintains its military forces to address what it perceives as its strategic circumstances. The militaries of France and the UK look most like the U.S. military. Neither nation faces any internal threats that require a military response, and, thus, their militaries are used abroad to pursue national policies and priorities. Both militaries deploy their forces overseas, but these deployments are limited to fit the size of the country's force and budget. Furthermore, deployed French and British forces often serve in a supporting role (e.g., contributing to coalition operations in Iraq or Afghanistan). Both France and the UK also employ significant TAA missions to extend their influence.

China and India, on the other hand, are focused on external threats and internal issues. They participate in few deployments, and those in which they do participate are almost exclusively noncombat operations conducted under the auspices of the United Nations (UN).

Finally, Israel faces a strategic circumstance that requires its armed forces to prepare for a mix of internal and external threats. Furthermore, these threats demand forces that are trained, organized, and equipped for high- and low-intensity operations and for contending with a state that does not share a border with Israel (i.e., Iran).

The militaries of the states we assessed are generally organized around general-purpose forces designed principally for combat operations. The UK, France, and Israel each visualize a range of operations that their forces may have to execute. Although they also rely on general-purpose forces, China and India mostly prepare their forces for a specific activity (e.g., COIN in India) or for operations relevant to a specific military challenge (e.g., a Taiwan contingency in China).

Additionally, China, France, India, and Israel all employ paramilitary specialty forces used for internal-security missions that lie somewhere between policing and military action (e.g., COIN, civil support, humanitarian assistance), and China has used its paramilitary forces to support a UN peacekeeping mission in Haiti. However, there does not appear to be a joint culture in any of these nations, except the UK.

All the nations we examined, except Israel and, to a degree, China, rely on volunteer forces (or are moving in that direction) and are reducing the size of their militaries. Long-serving volunteers are replacing conscripts, which results in higher costs but increased professionalism and a more sophisticated operational capability. Israel is the clear exception: There, universal service is still the basis of Israel's active-duty and reserve forces. In the active component, however, the Israeli system has resulted in a military without a noncommissioned-officer corps. Thus, Israeli junior officers pick up duties and responsibilities that, in other militaries, are in the realm of career noncommissioned officers.

Insights from Other Nations—Potential Best Practices

As already noted, each of the countries we examined relies on general-purpose forces organized principally for conventional combat operations. In this regard, the five nations are all very similar to the United States. There are, however, several differences evident in predeployment training, the use of subject-matter experts (SMEs), the approach to staff training, the use of combat training centers (CTCs), and approaches to the TAA mission. These differences, described in the following sections, may offer potential best practices for improving U.S. training systems.

Predeployment Training Can Build on Strong Traditional Skills

Training for traditional challenges appears to be highly successful in developing foundational individual and collective skills, skills that the British Army's Land Warfare Center calls the *adaptive foundation*. The term *adaptive foundation* refers to the starting point from which forces can subsequently be adapted to specific operational environments. The

UK's training and readiness cycle spends most of its time, and all of its CTC resources, preparing forces for traditional challenges. CTCs at Sennelager, Salisbury Plain, and the British Army Training Unit Suffield in Canada continue to focus on major combat operations (MCO), although they are integrating a more complex environment into training scenarios.

Preparing units for a specific operational environment (a process called *force generation*) requires a relatively modest commitment of time and resources, provided that units are well-trained in basic military operations (a process called *force preparation*). British forces have earned an enviable reputation at the level of the battle group and below in such diverse theaters as Northern Ireland, the Balkans, Sierra Leone, Afghanistan, and Iraq. The UK's Operational Training and Advisory Group (OPTAG) has the mission of preparing forces for deployment into these various theaters. OPTAG accomplishes this mission with fewer than 200 assigned military personnel (of very high quality). In a training and readiness cycle that lasts 24 months, OPTAG requires around one month to train the trainers; then, it allows the trainers to train their units and conducts a confirmatory exercise. India's various battle schools, including the XVth Corps Battle Schools, the Counter Insurgency and Jungle Warfare School, and the High Altitude Warfare School, prepare units in a similar fashion, represent a similarly modest commitment of resources and seem to prepare units well for asymmetric challenges.

SMEs Can Provide Crucial Capability
The training of units in the Indian Army is tailored to the specific region and specific operational conditions in which units are stationed. In day-to-day operations, this is a viable approach. When the units deploy to contingencies for which they have not prepared—as in the 1999 war in Kargil, examined in Chapter Five—this approach can prove inadequate. In the Kargil crisis, Indian troops acclimated and trained for tropical COIN operations were not prepared for conventional combat operations in the mountains. The insight from the Indian experience in Kargil is that a small group of SMEs—in this case, mountain-warfare experts—can rapidly infuse capability into units by enabling forces

trained for one environment or contingency to improve their performance in a different set of circumstances. In the United States, this SME approach could be a way to improve training for specific deployments or to improve the performance of units that are deployed against contingencies that were not the focus of predeployment preparations. To do this, the U.S. military should take advantage of SMEs for operations across the spectrum and for different types of complex terrain, whether mountainous, urban, or jungle. Furthermore, to leverage these SMEs, a system of identifying and tracking SMEs across the force has to be in place.

Staff Training Can Serve as a Vehicle to Prepare Forces for Multiple Contingencies

French processes for command and control (C2) training offer a potential model for training staffs for multiple types of operational contingencies. The effective transition of French forces in the Ivory Coast in 2004 from peacekeeping operations to irregular warfare[2] demonstrated very agile C2 capabilities and was certainly more effective than either the U.S. transition to stability operations in Iraq after MCO or the British response to the deteriorating situation in Basra from 2005 onward. The professional French response undoubtedly owed much to the force's highly unorthodox operational commander, General Henri Poncet, but it also points to the importance of highly trained staffs. French brigade staffs gain their proficiency by conducting three to four times as many command post exercises (CPXs) per year as either the U.S. or the British staffs. Furthermore, a number of these CPXs are externally evaluated.

The significantly greater frequency of CPX training in the French force is enabled in part by the fact that the French often train a single echelon at a very reduced scale. This training omits many of the ancil-

---

[2]   See U.S. Department of the Army, FM 3-0, *Operations*, Washington, D.C.: Headquarters, Department of the Army, 2008, p. 2-4. U.S. Army doctrine currently recognizes five major operational themes: peacetime military engagement, limited intervention, peace operations (of which peacekeeping operations are a subset), irregular warfare, and MCO. Of these, irregular warfare seems best to capture the cognitive thrust of what was going on in Côte d'Ivoire.

lary functions required to establish and maintain headquarters (HQ) and instead focuses narrowly on cognitive processes. Thus, conducting a meaningful CPX does not require coordinating the schedules of multiple HQ at several echelons—an effort whose scope, resource demands, and complexity deter frequent repetition. Technology also plays an enabling role. France's Simulation de Combat Interarmées pour la Préparation Interactive des Opérations for brigade HQ and above automates many of the entities, reducing the requirements for higher- and lower-control players.

The French process of increasing the proficiency of unit HQ seems to be highly effective in enabling units to master transitions, and their ability to do so shows that C2 training yields a high return on a marginal training investment. Thus, directed, evaluated CPXs could provide a training methodology for U.S. forces that could help address concerns, recently voiced by U.S. Army Chief of Staff General George Casey, about the deterioration of critical integration, synchronization, and other skills required to prevail across the full range of military operations. According to General Casey,

> Current operational requirements for forces and insufficient time between deployments require a focus on counterinsurgency training and equipping to the detriment of preparedness for the full range of military missions.[3]

CTCs Can Be Used Differently

Several of our case studies show that other countries believe that their training centers should mainly provide foundational combined-arms fire-and-maneuver training. The militaries build on these skills with predeployment training focused on the specific operational environment to which a given unit is deploying. We believe that reorienting U.S. training to a predeployment model along the lines of OPTAG or the Indian Army's Counter Insurgency and Jungle Warfare School would allow CTCs to return to a principal focus on task-force

---

[3] U.S. Department of the Army, *2008 Army Posture Statement: A Campaign Quality Army with Joint and Expeditionary Capabilities*, Washington, D.C.: Headquarters, Department of the Army, 2008, p. 6.

combined-arms fire-and-maneuver operations training. We believe that this training is critical to maintaining full-spectrum capabilities and to addressing General Casey's concerns. We are not implying that there should be a return to a "Fulda Gap" model; rather, we believe that the model employed by several of the countries we examined is worthy of close examination by the United States. Additionally, we are not advocating that the CTCs return to portraying a sterile battlefield. The British Army Training Unit Suffield has integrated villages, civilians, and other complications into its training scenarios. Similarly, the Israelis have a sophisticated urban-operations training facility in Tze'elim.

It is logical to assume that when units spend time at CTCs preparing for the operational environment in Iraq and Afghanistan, they are not using that time to train for synchronized brigade and battalion task-force combat operations. It makes eminent sense to prepare units for the specific operational context they will face, but doing so at a CTC sacrifices an opportunity to conduct foundational combined-arms training at facilities uniquely suited to provide this training. Contextual training could likely be done elsewhere (at lower cost) or become one component of the CTC experience. Clearly, U.S. CTCs are the locations best prepared to provide combined-arms training in intense simulated combat. And, as the Israeli experience in Lebanon in 2006 shows, intense combat is not so much about scale (i.e., battalion or brigade force-on-force engagements) as about the qualitative challenges hybrid adversaries can pose. Opponents with a modicum of training, organization, and advanced weaponry—like Hezbollah—create tactical and operational dilemmas that demand combined-arms fire and maneuver. Thus, based on their experiences in Lebanon in 2006, the Israelis have reoriented the focus of much of their training—particularly the training conduced at the Tze'elim training center—on what they call *high-intensity conflict* (HIC). Their subsequent performance in Gaza in December 2008–January 2009 seems to show that this reorientation was wise.

Moreover, because the goal of the training centers in France, the UK, and Israel is foundational rather than finishing, U.S. units might profitably undergo their CTC rotation earlier in their training cycle.

The point of maneuver exercises in the UK and Israel is as much to teach staff operations, planning, troop-leading procedures, and basic tactical skills as to teach specific collective tasks. British forces undergo CTC rotations toward the middle of their readiness cycle, and such rotations constitute most of French forces' collective preparations for an operational tour.

Approaches to the TAA Mission—the French and British Models

Of all the countries we assessed in this study, it appears that France and the UK have the TAA models that provide insights into improving the U.S. model. All three countries view TAA and building partner capacity (BPC) as ways in which they can favorably shape and influence the global security environment. That said, their TAA approaches differ significantly in several key areas: trainer selection, mode of deployment, training of the trainers, and career implications for the trainer.

In the United States and the UK, the processes for selecting trainers and advisers from the conventional forces do not appear to be particularly rigorous, and these assignments are not generally sought by officers in these two countries. What appears most different in the French model for selecting advisers is that service on advisory duty is expected of French officers who are competitive for advancement.

The U.S. system for preparing trainers and advisers emphasizes operational and tactical training over cultural training, and the cultural training that is available does not address key points, such as empathy with the advised, addressed in the French and British models. Although the French and British predeployment training for advisers lasts only about two weeks, the process appears to do a good job of ensuring that advisers are adequately trained. In the U.S. system, training for TAA lasts between two and six months.

There are no foreign-area officer programs in France and the UK; most of the forces deployed on TAA missions come from the pool of general-purpose forces and are generalists. France and the UK employ similar TAA models: Advisors are embedded with the partner, and they often wear the host-nation uniform. In the U.S. system, advisers have typically not been embedded in partner units, although this is happening now in Afghanistan, Iraq, and the Philippines (as it did during the Vietnam War), and embedding may evolve into the new norm. In the French system, the TAA mission is part of a deployed battalion's normal mission; advisory duty is part of the normal career path, and success in TAA missions is seen as a prerequisite for advancement. This is not the case in the UK or the United States. In the UK, TAA missions are encouraged but not necessarily career-enhancing. In the United States, TAA missions have traditionally not been part of mainstream career paths. Indeed, in the United States, advisory duty has generally been viewed as detrimental to advancement—it was what happened to an officer who was not competitive for more-important, career-enhancing assignments. Clearly, the importance of training the military forces of Iraq and Afghanistan as components of a successful strategy is understood within the U.S. military, and a recent message by General Casey stresses the importance of service on Training Teams and Provincial Reconstruction Teams.[4]

What Should OSD Do About These Insights?

Several overarching insights from our analysis lead to specific recommendations for OSD to pursue to improve current U.S. training practices. These insights and recommendations are in four areas: adapting to irregular challenges, preparing the force, defining TAA requirements, and preparing for future challenges.

Adapting to Irregular Challenges

The sponsor asked that we examine approaches to training forces to adapt to irregular challenges. There is an emerging literature that emphasizes the importance of individual and unit adaptability and, thus, improving methods to train both to be adaptable. Proponents of this training approach argue that

---

[4] George Casey, "CSA Sends—Transition Team Commanders (Unclassified)," June 17, 2008.

- The United States faces future threats that are irregular and asymmetric.
- Adaptability is key to meeting these challenges.
- Adaptability (and intuition) can be taught.

This adaptation tautology implies that fundamental change across the doctrine, organization, training, materiel, leader development and education, personnel, and facilities spectrum is not necessary to prepare for irregular and asymmetric challenges—well-trained individuals and units can adapt to any circumstance.

Although we generally believe this approach to individuals and units is important, in our view, it is necessary but not sufficient. Our opinion is that it is the role of the institutions within the Department of Defense (DoD) to prepare U.S. forces for the challenges that they will encounter in specific irregular (and regular) operations. The responsibility for adaptation must also belong to these institutions rather than to individuals and units. This is not to say that teaching critical thinking, decentralizing decisionmaking, and a host of other initiatives are not useful approaches. They are necessary but not sufficient, and they have always been valued, at least in theory, in the past.

That said, the important role of institutions is to provide an appropriate problem-solving framework for use by individuals and units when asymmetries present challenges that existing methods do not address adequately. Perhaps the best recent example of U.S. military institutions adapting themselves to new conditions is the case of the U.S. Army revising its fundamental concept about how to succeed in war. The 2001 version of Field Manual (FM) 3-0, *Operations*, posited a construct for warfare that had endured in the U.S. Army for nearly 80 years:

> The offense is the decisive form of war. Offensive operations aim
> to destroy or defeat an enemy. Their purpose is to impose US will
> on the enemy and achieve decisive victory.[5]

---

[5]   U.S Department of the Army, FM 3-0, pp. vii, 7-2.

This was the doctrine that the U.S. Army—a very well-trained and well-equipped force—took into Operation Iraqi Freedom, and, by 2006, it was clear that this approach was not adequate to deal with the insurgency that developed after the end of MCO. Eventually, the U.S. Army revised its approach, publishing, in conjunction with the U.S. Marine Corps, a new COIN manual, FM 3-24/Marine Corps Warfighting Publication (MCWP) 3-33.5, *Counterinsurgency Field Manual*, that fundamentally changed the basic construct for successful operations, noting that

> the cornerstone of any COIN effort is establishing security for the civilian populace. . . . Soldiers and Marines help establish HN [host nation] institutions that sustain that legal regime, including police forces, court systems and penal facilities.[6]

This institutional adaptation was a precondition for the increasingly successful COIN operations that followed the promulgation of the new doctrine. Quite simply, absent FM 3-24/MCWP 3-33.5, *Counterinsurgency Field Manual*, even the most-adaptable individuals and units were not able to solve the COIN problem across Iraq using FM 3-0, *Operations*.

Another difference between U.S. methods and those of several of the countries we examined is one of training focus for individuals and units. In France and the UK, training emphasizes building location-specific expertise as a means of adapting the overall force to the specific contingency. The French and the British—like the Americans—have also created the capacity to quickly infuse lessons learned from ongoing operations into the training for those preparing to deploy. This is done to adapt their militaries to operate in the places to which they are about to deploy and to train individuals within this specific context. This is different from trying to teach adaptability. It is more along the lines of creating deep, vicarious intuition by expanding patterns in training that can be recognized and referred to during operations. The key for the institution is to minimize how long any operational

---

[6]   U.S. Department of the Army and U.S. Marine Corps, FM 3-24/MCWP 3-33.5, *Counterinsurgency Field Manual*, Chicago: The University of Chicago Press, 2007, p. 42.

environment remains asymmetric. Thus, our sense is that adaptability is an institutional—not an individual—responsibility. The challenge in training individuals is to prepare them as much as possible for the specific environment of the future deployment.

Nevertheless, there appears to be a need to understand how to identify how individuals respond to complex situations when they are under pressure and when traditional hierarchical chains of command are unavailable to support decisionmaking. This seems particularly important in the case of advisers. Thus, although adaptability may not be a trainable trait, it might be a discriminator for key positions in which the ability to cope with uncertainty is important. That said, our sense is that more empirical investigation is needed to understand the potential of training individuals and units to be more adaptable. Our recommendations are as follows:

- OSD should support further empirical research to determine if adaptability can in fact be trained and if an individual's ability to adapt can be determined.
- If adaptability can be assessed and trained, OSD should establish processes to determine which assignments (e.g., advisory assignments) require adaptability.

Preparing the Force

There are several gaps in current processes for preparing the U.S. armed forces for the irregular—and regular—challenges they face. There are multiple populations to prepare. Nevertheless, our sense is that the greatest gap exists at the senior levels. Quite simply, there has never been a deeply substantive or rigorous system of continuing training or education for officers beyond their attendance at a senior-service college at the O-5 or O-6 levels. A number of the nations we examined recognize the need for continuing education beyond that provided by their equivalent of the U.S. senior-service college. The British have a higher-command and staff course, and the Israelis have a course for colonels, brigadier generals, and new division commanders. Because senior U.S. officers are responsible for preparing their units for the challenges of the future and for guiding their training, it seems important to provide them with continuing education. Our recommendation is as follows:

- OSD should assess current programs for the continuing training and education of senior leaders and recommend corrective action.

Defining TAA Requirements

There is currently no enterprise-wide system within the DoD or other U.S. government agencies to identify and prepare American officers for advisory or foreign-military training assignments. These assignments are generally conducted on a one-off basis and are not career-enhancing. Finally, there is no DoD-wide repository for best practices or lessons learned for these missions. Our recommendations are as follows:

- OSD should work with the Joint Staff and the U.S. military services to set standards for advisers (including selection criteria) and craft directives that ensure that adviser training assignments are career-enhancing. These efforts could be similar to measures taken after Goldwater-Nichols to ensure that joint duty became a viable assignment.
- OSD should create processes to capture and disseminate TAA- and BPC-specific best practices from across the U.S. government and from relevant foreign governments.

Preparing for Future Challenges

One of the central ironies about adapting to and preparing for irregular challenges is that such challenges then become the new "regular" challenges. Israel's performance during the 2006 Second Lebanon War is instructive in this regard. After years of adapting to the challenges of the intifadas, the Israeli Army, despite its competence in addressing what its doctrine calls *low-intensity conflict* threats, found itself not competent to fight the HIC it encountered in Lebanon. There, the Israel Defense Forces (IDF) faced an opponent that was qualitatively different than the opponent Israel had focused on for years. Hezbollah was a trained militia with modern weapons in prepared defensive positions.

Currently, the U.S. armed forces may be in a condition similar to that of the IDF in 2006. Multiple combat tours in Iraq and Afghanistan have created U.S. units and individuals with deep experience in COIN. Additionally, adapting to the significant demands of the operational environments in these active theaters of war has, not surprisingly, resulted in a diminishment of high-end combat skills among U.S. forces. Thus, the extraordinary proficiency of the U.S. force at doing what it is having to do now may in fact be diminishing its capacity—as it did with the IDF—to do something it might have to do in the future. In short, the U.S. military, particularly its ground forces, has lost some of its full-spectrum capability.

Several of the nations we examined have developed training regimes that assist them in adapting and preparing their units for different operational scenarios. India deployed SMEs to improve unit performance, the French use multiple and evaluated CPXs involving differing scenarios to prepare their HQ, and the British "train the trainer" for several deployment scenarios through their OPTAG process. All of these practices offer promise to improve the current U.S. training system. Our recommendations are as follows:

- OSD should support an analysis to determine which Universal Joint Task List tasks are atrophying.
- OSD should further assess CPX strategies that train and evaluate HQ for the full spectrum of operations, and it should support the development of exercises that allow staffs to maintain full-spectrum proficiency.
- OSD should assess the potential of SME training and devise processes to identify and track SMEs.

Final Thoughts

During our research, we found that the U.S. military is the source of best practices in many areas in every country we examined. Nevertheless, the processes used to attain full-spectrum capabilities in several of the nations we analyzed differ from those used in the United States, and they appear to work. Thus, as we believe we demonstrate in this monograph, there are areas in which the U.S. military can learn from the experiences of these other nations to improve its ability to perform more effectively across the range of military operations.

Acknowledgments

Many individuals helped us understand how the nations analyzed in this monograph approach training for full-spectrum operations. We are grateful to the members, past and present, of the French, British, Indian, Israeli, and U.S. militaries for the candid comments and advice they provided during our field interviews. Our sponsors, Frank DiGiovanni and Colonel Joseph Thome, were particularly helpful in shaping the direction of our research as our analysis evolved. Finally, Russell Glenn and Thomas Manacapilli provided very thoughtful and useful reviews of draft versions. This monograph is better for all of their efforts.

Abbreviations

1 ID	1st Infantry Division
ABACUS	Advanced Battlefield Computer Simulations
ACSSU	Air Combat Service Support Unit
ACSU	Air Combat Support Unit
AFSOUTH	Allied Forces South
APOD	air point of departure
BATUS	British Army Training Unit Suffield
BCPO	Bureau de Conduite de la Préparation Opérationnelle
BCTP	Battle Command Training Program
BGTU	Battle Group Training Unit
BMATT	British Military Advisory and Training Team
BPC	building partner capacity
C2	command and control
C2SD	Centre d'Études en Sciences Sociales de la Défense
CAFTT	Coalition Air Force Transition Team
CALL	Center for Army Lessons Learned
CAST	Command and Staff Trainer

CATT	Combined Arms Tactical Trainer
CDEF	Centre de Doctrine d'Emploi des Forces
CDS	chief of the defence staff
CEB	Centre d'Entraînement de Brigade
CEC	Centre d'Entraînement Commando
CEITO	Centre d'Entraînement de l'Infanterie au Tir Opérationnel
CENTAC	Centre d'Entraînement au Combat
CENZUB	Centre d'Entraînement aux Actions en Zone Urbaine
CEPC	Centre d'Entraînement des Postes de Commandement
CFAT	Commandement de la Force d'Action Terrestre
CFI	Controle Formation Individuelle
CIECM	Centre d'Instruction et d'Entraînement au Combat en Montagne
CJCSM	Chairman of the Joint Chiefs of Staff Manual
CJO	Commander Joint Operations
CMC	Central Military Commission
CNAM	Centre National d'Aguerrissement en Montagne
CNCIA	Commission Nationale du Contrôle Interarmes
COCOM	Combatant Commanders/Combatant Commands
CoFAT	Commandement de Formation d'Armée de Terre
COIN	counterinsurgency

CPCO	Centre de Planification et de Conduite des Opérations
CPF	Centre de Préparation des Forces
CPX	command post exercise
CT	collective training
CTC	combat training center
DCT	Directed Continuation Training
DFID	Department for International Development
DOB	deployed operational base
DoD	Department of Defense
DOTMLPF	doctrine, organization, training, materiel, leader development and education, personnel, and facilities
EAW	Expeditionary Air Wing
ECOWAS	Economic Community of West African States
EMA	État-Major des Armées
EMIA-FE	État-Major Interarmées–de Force et d'Entraînement
EMSOME	École Militaire de Spécialisation de l'Outre-Mer et de l'Étranger
ESM	École Spéciale Militaire
EU	European Union
FAO	foreign-area officer
FCO	Foreign and Commonwealth Office
FHQ	force headquarters
FM	field manual
FMO	Force Multinationale d'Observateur

FORM	Force Operations and Readiness Mechanism
FOST	Flag Officer Sea Training
FTX	field training exercise
GAO	Government Accountability Office
GDP	gross domestic product
HIC	high-intensity conflict
HQ	headquarters
IAF	Israeli Air Force
ICS	Interagency Coordination Symposium
ID	infantry division
IDA	Institute for Defense Analyses
IDF	Israel Defense Forces
ITEA	Interagency Transformation, Education and Analysis
JANUS	Joint Army Navy Uniform Simulation
JFCOM	U.S. Joint Forces Command
JFHQ	joint-force HQ
JFSC	Joint Forces Staff College
JIACG	Joint Interagency Coordination Group
JMC	joint maritime course
JNTC	Joint National Training Capability
JP	joint publication
JRRF	Joint Rapid Reaction Force
JSOU	Joint Special Operations University

JTF	joint task force
JTFC	joint task-force commander
JTFX	joint task-force exercise
JTS	Joint Training System
LIC	low-intensity conflict
MCO	major combat operations
MCTAG	Marine Corps Training and Advisory Group
MCWP	Marine Corps Warfighting Publication
METL	Mission-Essential Task List
MiTT	Military Transition Team
MOB	main operating base
MOD	Ministry of Defence
MOU	memorandum of understanding
MP	Member of Parliament
MR	military region
MRAF	MR Air Force
NATO	North Atlantic Treaty Organization
NCO	noncommissioned officer
ONU	Organisation des Nations Unies
OPTAG	Operational Training and Advisory Group
OSD	Office of the Secretary of Defense
OST	Operational Sea Training
OTIADEX	Exercice d'Organisation Territoriale Interarmées de Défense

OTRI	Operational Theory Research Institute
PDT	predeployment training
PJHQ	Permanent Joint HQ
PLA	People's Liberation Army
PLAAF	PLA Air Force
PLAN	PLA Navy
PLANAF	PLAN Air Force
RAF	Royal Air Force
RCA	Regiment Chasseurs d'Afrique
RCT	Requested Continuation Training
RECAMP	Reinforcement of African Capabilities to Maintain Peace
RN	Royal Navy
ROE	rules of engagement
ROTC	Reserve Officers' Training Corps
RPG	rocket-propelled grenade
SAR	safety and readiness
SCETC	Security Cooperation Education and Training Center
SCIPIO	Simulation de Combat Interarmées pour la Préparation Interactive des Opérations
SDR	Strategic Defence Review
SFA	security force assistance
SME	subject-matter expert
SOCOM	U.S. Special Operations Command

SOD	systemic operational design
SOF-IA	Special Operations Forces–Interagency Collaboration Course
SOS	special operations squadron
SPT	support
T2	training transformation
TAA	train, advise, and assist
TAAF	Terres Australes et Antartiques Françaises
TMAAG	Theater Military Advisory and Assistance Group
TRSS	Terrorism Response Senior Seminar
UAE	United Arab Emirates
UAV	unmanned aerial vehicle
UE	Unified Endeavor
UJTL	Universal Joint Task List
UN	United Nations
USMC	U.S. Marine Corps

CHAPTER ONE

Introduction

Purpose

The origins of this study lie in an attempt to answer a rather straight-forward, but thus far largely unanswered, question: What can the U.S. military learn from other militaries about how better to prepare for full-spectrum operations and deployments? To this end, RAND was asked by the Office of the Secretary of Defense (OSD) for Personnel and Readiness to examine the militaries of China, France, the UK, India, and Israel.

Background

The challenges the U.S. Department of Defense (DoD) faces now and will face in the future will require the department to maintain capabilities to conduct the full range of military operations and a broad set of military missions, which are delineated in Joint Publication 3-0, *Joint Operations*, and shown in Figure 1.1. We believe that these challenges necessitate the exploration of ways to make U.S. military training more relevant to existing and emerging challenges.

The difficult and continually evolving operations in Iraq and Afghanistan, for example, show the complexities of what is now termed *irregular warfare* and highlight the need for new approaches to the security challenges with which the United States is now contending and will likely confront in the future. However, as this monograph relates, other nations have experience in preparing for full-spectrum challenges.

Figure 1.1
The Range and Types of Military Operations

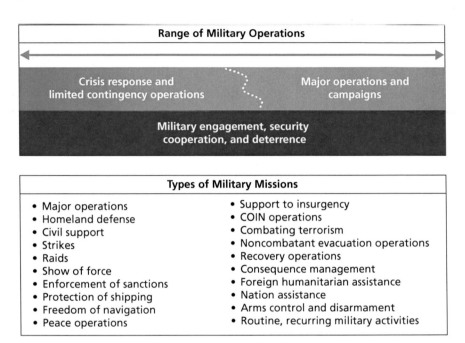

SOURCE: U.S. Joint Chiefs of Staff, Joint Publication 3-0, *Joint Operations*, 2006, pp. I-12–I-13.
RAND *MG836-1.1*

Tasks

At the outset of the project, the RAND team, in coordination with the study sponsor, developed three tasks:

- **Task 1: Review foreign training objectives, models, methodologies, and applications.** Review unclassified and classified reports. Consult with subject matter experts (SMEs) in the United States and in partner countries, where appropriate. Review available after-action reports and intelligence assessments of foreign collective training for small infantry units and their operational application, where possible. Highlight key aspects of indigenous training approaches (e.g., emphasis on certain types of conflicts, use of innovative methods), including cultural aspects. Identify ways in which the training influenced the conduct of subsequent operations.

- **Task 2: Identify comparable U.S. training practices relative to foreign militaries.** Review DoD guidance and unclassified and classified reports, including the DoD Training Transformation (T2) Implementation Plan. Consult with SMEs in the United States to identify common training practices for collective training of U.S. small infantry units. Focus this task on comparable training practices identified during Task 1.

- **Task 3: Analyze similarities and differences in approaches.** Compare the military training practices identified during Task 2. Highlight any obvious differences in training approaches. Recommend appropriate objectives, models, methodologies, and practices that warrant further study for possible incorporation into U.S. training processes to improve effectiveness and operational readiness.

These tasks evolved over the life of the project. Specifically, the study was expanded to investigate, to the degree that sources were available, training and readiness methodologies in the ground, air, naval, and specialty forces of the countries we examined. Additionally, we assessed how these countries try to adapt their forces for operations across the spectrum of operations.

Additional Tasks

We also considered two areas not within the original scope of the study: (1) how the assessed nations approach the issue of adaptability and (2) the train, advise, and assist (TAA) mission.

Adaptability
In assessing the degree to which each nation's training model imparts adaptability, we found that the Institute for Defense Analyses (IDA)

study called *Learning to Adapt to Asymmetric Threats* provided a useful framework for analysis. The IDA study noted that successfully inculcating adaptability neither can nor should be an exclusive function of individual training, professional development, or collective training; rather, such training should span and integrate all three domains to develop an effective, adaptable force. To that end, the study enumerated three broad imperatives:

- **Train for adaptability.** Adaptability is an acquired skill at both the individual and collective levels. To acquire it, individuals or teams should undergo repetitive training in unfamiliar and constantly changing scenarios, which will require them to adapt. For instance, a command-post exercise (CPX) might iterate training on a single problem under rapidly changing conditions.
- **Teach cognitive skills.** Adaptability requires individuals to develop their intuition and their critical- and creative-thinking skills. Extensive experience and frequent repetition under frequently and rapidly changing circumstances contribute to developing individual intuition. The IDA authors do not prescribe particular methods for developing these skills, but we assume high-quality academic instruction is needed to contribute to their development. Because the inculcation of these attributes is primarily a matter of leader development and individual training, an understanding of a given country's relative emphasis on leader development and individual training is critical to understanding the institutional context in which collective training takes place.
- **Develop relational skills.** Colloquially, this is a matter of teaching people to "play well with others." It requires the development of individual skills (especially self-awareness) and team skills, such as social awareness and relationship management.[1]

The study further identifies three key audiences: individuals, commander-leader teams, and units. Thus, to the extent that leader devel-

---

[1]   John C. F. Tillson, Waldo D. Freeman, William R. Burns, John E. Michel, Jack A. LeCuyer, Robert H. Scales, and D. Robert Worley, *Learning to Adapt to Asymmetric Threats*, Alexandria, Va.: Institute for Defense Analyses, D-3114, 2005, pp. 39–54.

opment and collective training broadly align with these imperatives, we can say that the training model emphasizes inculcating adaptability.

According to the IDA study, the U.S. training model does not yet embrace adaptability; rather, it trains individuals and units to cope with known asymmetries by applying proven techniques to teach new skills.

Train, Advise, and Assist: The U.S. Military Approach to Predeployment Training for TAA Missions

To draw insights that might be useful in the U.S. system, we assessed how France and the UK address the TAA mission. To set the stage for that assessment, we provide a description of the U.S. system.

Despite having been involved in TAA missions for many decades, the DoD, since the Vietnam War, has not tended to prioritize predeployment training for TAA missions. Only within the last several years has the DoD begun to initiate new programs designed to improve the U.S. military's proficiency in working with a variety of interagency and coalition partners. At the joint level, several new training and exercise programs have been created to address this deficiency. However, the new programs share no common, standard approach. The military departments, moreover, have been slower to respond to the deficiency. Since 2006, apparently after recognizing some of the lessons from U.S. training experiences in Iraq and Afghanistan, the U.S. Army, the U.S. Marine Corps (USMC), and the U.S. Air Force have initiated some training programs for this niche (vice core) mission. The two sections that follow provide an overview of some of the key joint and service-level predeployment training initiatives as a baseline for comparison with the approaches employed by France and the UK.

Joint-Level Predeployment Training for TAA Missions.[2] In this section, we outline some of the key training activities conducted at the joint level for TAA missions.

---

[2] This section on joint predeployment training contains excerpts from Michael Spirtas, Jennifer D. P. Moroney, Harry J. Thie, Joe Hogler, and Thomas Durrell-Young, *Department of Defense Training for Operations with Interagency, Multinational, and Coalition Partners*, Santa Monica, Calf.: RAND Corporation, MG-707-OSD, 2008. Note that new training

The National Defense University Interagency Coordination Symposium. The Interagency Transformation, Education and Analysis (ITEA) program at the National Defense University conducts the Interagency Coordination Symposium (ICS). The ICS is a strategic-level seminar aimed at Joint Interagency Coordination Group (JIACG) representatives and joint staff officers. The seminar is held once per quarter in Washington, D.C., and a tailored, on-site version can be provided when requested by Combatant Commanders (COCOMs). The course normally lasts two to three days when conducted on site and a full week when held at the National Defense University.

ITEA ICS program coordinators stated that, in addition to providing valuable interagency training, the program is helpful in informing the debate about what kinds of skills an integrated operator should possess and what corresponding training curriculum should be developed.[3]

ITEA seminar presenters are typically brought in from various outside agencies and departments to supply relevant interagency views and lessons from experts in the field. Individuals with JIACG experience are often asked to lecture at the seminars. Blocks of instruction within the symposium include

- Interagency Coordination Overview: The National Security Council and the Homeland Security Council
- Coordination with State and Local Governments
- Coordination Challenges Case Studies/Lessons Learned
- U.S. Government Capabilities and Coordination Exercise.[4]

The Joint Special Operations University (JSOU). The JSOU offers two courses specifically relevant to TAA preparation.

---

centers are being created to improve performance in Iraq and Afghanistan. One example is the COIN Center for Excellence in Camp Taji, Iraq.

[3] Discussions with ITEA program coordinators, Washington, D.C., spring 2006 and January 2008.

[4] Interagency Transformation, Education and Analysis, "ITEA Interagency Coordination Symposium Agenda," December 12–14, 2006.

The Special Operations Forces–Interagency Collaboration Course (SOF-IA). SOF-IA is conducted at the O-4 level and below. The course was developed in 2006 and had been taught several times by the end of 2007. SOF-IA focuses only on preparing military personnel to work with other U.S. governmental agencies and does not directly address working with multinational organizations or coalition partners. Relevant blocks of instruction include

- The National Security Council and the Interagency Process
- The Interagency Process
- Collaboration with Other Agencies
- The Embassy Country Team
- The JIACG/the Joint Interagency Task Force
- Collaboration with Intelligence Agencies
- Shaping the Environment: Security Assistance and Foreign Internal Defense
- The Special Operations Forces–Interagency Collaboration Exercise.[5]

The Terrorism Response Senior Seminar (TRSS). TRSS, developed in 1977, primarily targets special-operations personnel at the O-5 level and above. Federal Bureau of Investigation and Department of Energy representatives often attend, but there is typically no other interagency participation (with the exception of the invited speakers). Class size is normally 25–30 students, and 15 is the minimum. Presentations are made by guest speakers from various organizations within the interagency community. The JSOU provides the course objectives to the speakers in advance to help them prepare material and keep class discussions on track.

The course is offered several times per year, but few students participate. We had planned to observe a portion of TRSS during a visit to the JSOU, but the seminar was cancelled due to insufficient

---

[5]   Joint Special Operations University, "SOF-Interagency Collaboration Course (Planning Draft)," July 21, 2006.

interagency participation. When the course is held, relevant blocks of instruction include

- Interagency Roles
- Roles of Other U.S. Government Agencies
- Security Assistance and Foreign Internal Defense
- Terrorism-Response Case Studies.[6]

The Joint Forces Staff College (JFSC). The JFSC's Joint, Interagency, and Multinational Planners' Course is offered five times per year at U.S. Joint Forces Command (JFCOM); the class size is about 25 students. The course was taught for the first time in January 2006, and, since then, it has been conducted only three other times. The target audience for the course is the COCOM JIACG action officer, and the ideal mix of participants is 50 percent DoD and 50 percent non-DoD. However, the non-DoD participants generally only account for 30 percent of the course participants.[7] Blocks of instruction are taught primarily by guest lecturers, much as occurs in the ITEA and JSOU courses.

The focus of the course is on helping the student understand how National Security Council guidance affects COCOM planning. Relevant blocks of instruction include

- The Interagency Process
- Interagency Players in Complex Contingencies
- The Country Team
- The Ambassador/Country Team and the Military
- The JIACG Concept
- Intelligence Support to the Interagency

---

[6]   Joint Special Operations University, "Terrorism Response Senior Seminar Plan of Instruction," December 5–7, 2006; discussions with a course leader, Hurlburt Field, Fla., spring 2006 and January 2008.

[7]   Discussions with a course coordinator, Hurlburt Field, Fla., December 2006 and May 2008. Interagency representatives are usually country-team members serving abroad in U.S. embassies.

- Nongovernmental Organizations and Transnational Corporations
- Interagency Exercise Coherent Kluge.[8]

Recent reforms in transforming U.S. joint training have included placing importance on integrated operations. In accordance with the Unified Command Plan, JFCOM provides joint training expertise, inter alia, to the COCOMs and the services. This is achieved by supporting the development of joint training requirements and methods and supporting the execution of joint exercises. A key element of the joint-exercise program is the series of exercises that JFCOM organizes for each the geographic COCOMs. Typically, JFCOM organizes two to three mission-rehearsal exercises each year for each of the commanders. In addition, as part of the T2 initiative of the Secretary of Defense, JFCOM has become one of the key actors in bringing T2 approaches to joint training and exercises to joint task forces (JTFs). An essential tool in furthering T2 objectives and training COCOMs and JTF headquarters (HQ) has been JFCOM's singular mission of accreditation and certification of key capabilities of COCOM staffs. Thus, through a series of exercises organized by JFCOM, COCOMs and JTF HQ staffs are tested in critical areas identified as essential to mission success.[9]

One of the principal tools employed by JFCOM in supporting the joint training objectives of the COCOMs has been the Unified Endeavor (UE) exercise series. Begun in 1995, UE exercises allow JTF component commanders and their staffs to train at the operational level of war. Typically, UE exercises consist of a three-phase program that ends in a computer-aided exercise. The three exercise phases consist of academic training, development of operations plans, and execution of the operations orders.

Service-Level Perspectives. At the service level, several key programs are worth noting. This section outlines the steps taken by the

---

[8]   Joint Forces Staff College, "Joint Interagency Multinational Planner's Course," JIMPC 07-1, November 13–17, 2006.

[9]   U.S. Joint Chiefs of Staff, CJCSM 3500.03A, *Joint Training Manual for the Armed Forces of the United States*, Washington, D.C.: Joint Staff, 2002, pp. G1–G8.

U.S. Army, the USMC, and the U.S. Air Force to develop a more-institutionalized approach to predeployment training for TAA. Institutionalization, however, has not been the primary goal of U.S. military TAA training. Rather, the focus for the services has tended to be on standing up programs for this niche mission to respond to shortfalls in trainers in Iraq and Afghanistan. Since late 2007, however, there has been talk among senior service leadership of making TAA training permanent. Furthermore, a 2008 message from U.S. Army Chief of Staff George Casey stressed the importance of transition-team duty.[10]

The U.S. Army. The U.S. Army conducts one course that is relevant to TAA, and has considered a new concept for this mission, as discussed below.

Military Transition Team (MiTT) Training. To prepare U.S. personnel for the task of training the Iraqi military, the U.S. Army retasked elements of the 1st Infantry Division (ID) to enable the organization and its personnel to focus solely on MiTT training.[11] At Fort Riley, the 1 ID conducts training to prepare MiTTs for deployment to Iraq. MiTTs are typically composed of a ten-person team of mostly conventional forces. The teams are assigned to the Iraqi Assistance Group and are embedded in Iraqi units, mostly at the battalion and brigade levels. MiTTs at the brigade level typically include one special-operations officer (a major or a lieutenant colonel).

The 1 ID focuses its MiTT training largely on the tactical and technical skills the team will need to impart to its Iraqi counterparts. There is little emphasis on teaching these future MiTT members how to effectively transfer that knowledge. Specifically, during the 54-day program, 1 ID originally reserved just three afternoons for cultural awareness and three days for teaching and advising. The 1 ID devotes the remaining 50 days to teaching tactical skills. Several sessions on language training are provided, but students learn only survival lan-

---

[10] George Casey, "CSA Sends—Transition Team Commanders (Unclassified)," June 17, 2008.

[11] See Peter Spiegel, "Army Is Training Advisors for Iraq," *Los Angeles Times,* October 25, 2006.

guage skills.[12] Officers involved in the training have identified other shortcomings. For example, in July 2006, one former battalion commander wrote a critique of the program that emphasized the need for additional training in language skills, cultural awareness, negotiating, building rapport, and interpersonal skills.[13] According to the officer, MiTT members often fail to comprehend the vast cultural differences they will face. The cultural divide has occasionally caused some MiTT members to simply avoid interacting with their Iraqi counterparts. The obvious implication is that the mission is negatively affected.[14]

The Theater Military Advisory and Assistance Group (TMAAG). In an attempt to institutionalize the MiTT concept, the U.S. Army is considering the development of TMAAGs, which would provide the U.S. Army Service Component Commands with forces to execute specified theater-security cooperation tasks or activities in support of the COCOMs. Each TMAAG would consist of a small administrative HQ and three assigned training teams, each with approximately 22 personnel able to execute security-cooperation or TAA missions with host-nation militaries. Ideally, they would train host-nation forces, conduct detailed assessments of needs and capabilities, and facilitate participation of U.S. Army–provided rotational general-purpose forces.[15] In early 2009, the U.S. Army had not fully approved the TMAAG concept and had planned to study the idea further before making any decisions.

The USMC. Due to high demand for transition teams, the USMC created the Security Cooperation Education and Training Center (SCETC) in an effort to augment its ability to provide this critical capability and to standardize training procedures. In the past, individual Marine Expeditionary Forces conducted this training on their

---

[12] Telephone discussion with a U.S. Army lieutenant colonel (assigned to 1 ID, Fort Riley, Kans., and responsible for the conduct of MiTT training), May 2007.

[13] Telephone discussion with a U.S. Army major who was formerly assigned to an MiTT, May 2007.

[14] U.S. Department of the Army, "Military Transition Team OIF-OEF Training Model," Web page, undated.

[15] U.S. Department of the Army, "TMAAG Concept Paper," May 12, 2008.

own. In addition to providing MiTTs, the USMC is providing Border Transition Teams and National Police Training Teams, even though border transition and police training are not normally USMC mission areas. Some of the training is conducted at the units' home stations, and some is conducted at a newly constructed facility at the Marine Corps Air Ground Combat Center. Unlike the U.S. Army's MiTTs, the USMC MiTTs are typically composed of personnel from a single unit.

Like the U.S. Army's effort at Fort Riley, the USMC MiTT training focuses on the skills the teams will need to teach their foreign partners. SCETC also provides language and cultural training. It uses role-playing and has students act out scenarios, both of which help teach the transition teams how to work with people from other cultures. However, at the time of writing, the training did not include interaction with other U.S. government agencies, such as the Immigration and Customs Enforcement section of the Department of Homeland Security (which has worked with the U.S. Army's 1 ID), although this is a goal.

Presently, the USMC is working to institutionalize and strengthen its ability to train military advisers. It is divesting SCETC of its training responsibilities and is working to create the Marine Corps Training and Advisory Group (MCTAG), an independent command that will organize, train, and equip adviser teams.[16] The MCTAG concept was approved by the Commandant of the Marine Corps in October 2007. Initially conceived to address staffing shortfalls for TAA in Afghanistan and Iraq, the MCTAG's current purpose is to source the USMC's advisory capability to support mission requirements that exceed those in Iraq and Afghanistan. At the time of writing, MCTAG was only an embryonic capability, but it is expected to grow to constitute a cadre of trained advisers organized into regional branches and capable of being deployed in a scalable fashion.[17]

---

[16] U.S. Marine Corps, Marine Corps Training and Advisory Group, "U.S. Marine Corps Forces Command," briefing, October 2007.

[17] U.S. Marine Corps, Marine Corps Training and Advisory Group, "MCTAG Information Paper," May 12, 2008.

The U.S. Air Force. The U.S. Air Force conducts two courses that prepare its personnel for the TAA mission.

The Combat Aviation Advisory Course. The 6th Special Operations Squadron (SOS) at Hurlburt Field, Florida, conducts the Air Force Special Operations Command Combat Aviation Advisor Mission Qualification Course. The course seeks to prepare the squadron's personnel to serve as advisers and provide training to foreign air forces. These activities constitute the squadron's primary mission.[18] The course consists of four phases, and, like the transition-team training conducted by 1 ID and SCETC, it teaches a mix of technical skills and integrated-operations skills. Through a combination of lecture and practical application, the 6 SOS ensures that its members are versed in cross-cultural communications and integration techniques, regional studies, instructor and adviser techniques, security-assistance management, and interpreter and translator operations. The culmination of the second phase of training is the week-long Raven Claw Exercise, during which students apply approximately three months of academic training. The students' technical skills and ability to apply the integrated-operations skills they have learned are evaluated.

After completing the Raven Claw Exercise, the students complete two to four months of intensive language training before finally beginning the fourth phase, which consists entirely of technical aviation-related skills. Specific courses include

- Methods of Instruction
- Adviser Techniques
- Security-Assistance Management
- Political/Cultural-Integration Techniques
- Civil-Military Operations
- The Dynamics of International Terrorism
- Defense Security Assistance Management
- Contemporary Insurgent Warfare

---

[18] Telephone discussions with 6 SOS personnel, January 2007 and May 2008.

- Regional Orientation
- Cross-Cultural Communications.[19]

Coalition Air Force Training Team (CAFTT) and Combined Air Power Transition Force. For general-purpose forces, the U.S. Air Force has created a new course at Fort Dix, New Jersey, that includes elements taken from the 6 SOS Combat Aviation Advisory course and the U.S. Army's Fort Riley MiTT. This course, which prepares airmen for TAA missions in Iraq and Afghanistan, is 23 days long. Its focus is on training general-purpose forces in basic combat skills in a nonpermissive environment. Seven days of the course focus on training tactical and operational skills, and the remaining 16 days focus on cultural awareness. Advisers from 6 SOS deliver a seminar within the course. The course includes sessions on political Islam (which are taught by university professors), COIN, and cultural familiarization in Iraq and Afghanistan. Language training is conducted every day.[20]

More broadly, the U.S. Air Force aims to institutionalize TAA by creating Theater Air Advisory Squadrons. These squadrons could supply key units with embedded, trained advisers, or they could be assigned to air component commands. The benefit of the latter construct is that the squadrons could develop an area-of-responsibility focus over the long term. In this approach, a core cadre would be the framework of the squadron, and airmen with the desired skills could be drawn from forces serving in a normal deployment cycle.

In May 2008, OSD gave U.S. Special Operations Command (SOCOM) the lead role in developing the joint training and doctrine for security force assistance (SFA), broadly defined. At the time of writing, SOCOM's efforts were in the relatively early stages, but SFA has been incorporated into SOCOM's Global Synchronization Conference (held twice a year) as a working group. SOCOM was tasked to provide

---

[19] Air Force Special Operations Command, "AFSOC Combat Aviation Advisor Mission Qualification Course Formal Training Pipeline," undated. The Air Force Special Operations School also teaches a national aviation resource-development course. The 6 SOS sends its combat aviation advisors to this course.

[20] Telephone discussions with a CAFTT training specialist at the Air Education and Training Command, May 2008.

its ideas on the way ahead for SFA doctrine and training-curriculum development for general-purpose forces and to continue fleshing out the SFA mission requirements through the Global Synchronization Conference process. The SOCOM staff were, at the time of writing, actively soliciting feedback from the services and SMEs on the best way to proceed.

The U.S. Training System as a Baseline

Before beginning an analysis of foreign-military training models and methodologies, it is useful to understand the U.S. training model, which can provide a baseline for comparison. In explaining this model, we found it helpful to adopt two terms from the UK's doctrinal lexicon: *force preparation*, which means preparing forces for war in general; and *force generation*, which means preparing forces for *the* war. (See the brief discussion on definitions in Chapter Seven for fuller definitions of these and other terms.) The formal U.S. training model is the Joint Training System (JTS), which is defined in Chairman of the Joint Chiefs of Staff Manual (CJCSM) 3500.03A, *Joint Training Manual for the Armed Forces of the United States.*

The JTS is a force-preparation model in which U.S. forces are prepared for operations in general (albeit with a certain inclination toward operations in the area of responsibility of the COCOM to which the forces are assigned). In this system, COCOMs and the military services share responsibility for the training. COCOMs determine what tasks are to be trained, by whom, under what conditions, and to what standard, and the services plan and execute training to achieve these ends. Joint exercises and the Joint National Training Capability (JNTC) also serve to improve general proficiency in joint warfighting tasks. The Universal Joint Task List (UJTL), a key feature of the U.S. model, provides a common lexicon and a set of standards

The Defense Science Board has characterized the U.S. system as a "combat training center" model because of the central role combat training centers (CTCs), such as the U.S. Air Force's Red Flag Exercise and the U.S. Army's National Training Center, play in training.[21] CTCs, which attempt to replicate battles and engagements under extremely realistic conditions, are a central feature of the U.S. system.

Under the pressure of structural changes, however, the DoD is, in practice, moving toward a force-generation model. Structural changes include service adoption of modular force packages and cyclic readiness, the DoD's adoption of the Global Force Management construct, and the need to meet the current high demand for forces. Force generation, with which our research was primarily concerned, tends to be mostly a service responsibility, especially in practice. Generally, the services are better able to forecast which of their forces will be allocated against a COCOM's requirements than are either the gaining COCOM or JFCOM. Especially under current conditions, forces are often not formally assigned to gaining COCOMs until just prior to deployment, making it difficult for the COCOMs to lead effective force generation.

Force Preparation: A Shared Responsibility

The formal U.S. JTS extends U.S. Army training doctrine developed during the 1970s and 1980s for the Cold War era across the entire DoD. It assumes a certain degree of uncertainty about where, when, and under what conditions specific forces may be employed. It also assumes that U.S. forces will be called upon to perform a fairly narrow range of strictly military tasks (i.e., combat operations oriented on defeating or destroying conventional forces whose organization, training, and equipment are broadly symmetrical with those of U.S. forces). How these tasks are carried out may change slightly as geographic conditions vary, but their fundamental nature, and the effects sought through their application, are assumed to remain constant. These

---

[21] Joe Braddock and Ralph Chatham, *Report of the Defense Science Board Task Force on Training Superiority and Training Surprise*, Washington, D.C.: Office of the Under Secretary of Defense for Acquisition, Technology & Logistics, 2001, p. 7.

assumptions permit the creation and implementation of a fairly generic training regimen for all U.S. forces.[22]

U.S. policy and doctrine make force preparation a shared responsibility. COCOMs determine the tasks to be trained, the audience to be trained, the conditions under which training occurs, and the standards to which the audience is to be trained. They are also responsible for the planning and conduct of joint exercises in their area of responsibility. The military services and the combat support agencies (e.g., the Defense Intelligence Agency) plan and conduct training in accordance with these requirements.[23] Based on an assessment of likely missions, COCOM staffs then identify the relevant UJTL tasks to be taught. UJTL tasks range from the strategic, national level (a task called "Maintain Global Strategic Military Information and Force Status" is one example) to the tactical level (a task called "Employ Firepower" is one example).

Services are then responsible for preparing forces and HQ to perform these UJTL tasks and subordinate service tasks. The centerpiece of service training efforts has been rotations at the CTCs, such as the National Training Center at Fort Irwin, California, or the Air Force CTC at Nellis Air Force Base, Nevada. The U.S. Army's Battle Command Training Program is a CTC for division and corps staffs. COCOMs also conduct joint exercises to improve general joint warfighting skills. Joint exercises usually emphasize constituent CPXs; actual maneuver forces are used to a lesser extent because of the prohib-

---

[22] For example, the methodology for developing mission-essential tasks described in U.S. Joint Chiefs of Staff, CJCSM 3500.03A, seems to be derived from U.S. Department of the Army, FM 25-100, *Training the Force*, Washington, D.C.: Headquarters, Department of the Army, 1998.

[23] U.S. Joint Chiefs of Staff, CJCSI 3500.01C, *Joint Training Policy and Guidance for the Armed Forces*, Washington, D.C.: Joint Staff, 2004, Enclosure A, Introduction, paras. 5.b.–5.c. COCOMs derive potential missions from various sources, including the National Security Strategy, the Defense Strategy, the National Military Strategy, Security Cooperation Guidance, Contingency Planning Guidance, and plans that COCOMs develop as part of the Joint Strategic Capabilities Plan. The last two sources, which are classified, provide definitive guidance about which contingencies COCOMs must prepare to execute. Nonetheless, based on their own analysis, COCOMs may identify other missions as being implicit in strategic guidance from other sources.

itive expense of conducting large-scale maneuvers. To prevent adversaries from deducing U.S. war plans, joint exercises usually focus on generic warfighting skills rather than on a specific contingency.

In practice, service needs and priorities have tended to dominate even the joint aspects of the JTS. This is because the services provide units to the COCOMs that have been trained, organized, and equipped in accordance with individual service standards. Thus, the Government Accountability Office's (GAO's) finding that most joint training events have tended to focus at the COCOM level is not particularly surprising. Few training events involving actual forces (even forces from more than one service) can be accurately characterized as *joint*. The GAO noted that, because this condition has persisted during wartime, many units arriving in theater required additional joint training to prepare them for the operations they would have to execute. There is no joint equivalent to the U.S. Army's Battle Command Training Program (BCTP).[24]

The DoD is attempting to enhance joint training with the JNTC, a construct in which the department plans to invest almost $2 billion by 2010. The idea is to link service CTCs with information technology and then integrate training within a joint context. JFCOM has the lead for this effort, and the GAO has identified some initial successes. Like the early service training centers, however, the JNTC is focused more on generic force preparation than on force generation. Indeed, the GAO faulted the JNTC for focusing on conventional combat operations to the exclusion of emerging threats.[25]

In summary, the JTS deals with force preparation—enhancing the general capability of organizations to perform generic military tasks in a joint framework.

---

[24] See U.S. Government Accountability Office, *Military Training: Actions Needed to Enhance DoD's Program to Transform Joint Training*, GAO-05-548, Washington, D.C., 2005, pp. 3–4.

[25] U.S. Government Accountability Office, *Management Actions Needed to Enhance DOD's Investment in the Joint National Training Capability*, GAO-06-802, Washington, D.C., August 2006.

Distinctive Features of the U.S. Training Model

An extensive articulation of tasks—including through the UJTL and subordinate service task lists—is one of the distinctive features of the U.S. training model. The UJTL is over 800 pages long and, as previously noted, identifies and describes tasks ranging from the strategic, national level to the tactical level. These descriptions include the task, the general conditions under which it is to be taught, and the training standard to be met. A level below the UJTL are service task lists that, in theory, further decompose joint tasks into their service-unique components. Overall, this detailed and extensive articulation of tasks at all levels seems to reflect a doctrinal worldview peculiar to the U.S. armed forces: that the entire universe of possible actions required in warfare can be anticipated, that the role of units and individuals is to execute these tasks as defined, and that the role of leaders is to ensure compliance with these models.

CTCs are the other unique feature of the U.S. model. At these centers, U.S. forces perform simulated engagements and battles against dedicated opposing forces. Sophisticated engagement systems replicate the effects of both direct and indirect fires, and a professional staff helps units assess their performance and remedy deficiencies. Indeed, a group convened in 1999–2000 by the National Intelligence Officer for Conventional Military Issues concluded that the high cost of establishing such training centers effectively denied U.S. adversaries the ability to improve their proficiency using this model. Although training at these CTCs theoretically addresses units' mission-essential tasks, which vary by unit, in practice, the CTCs' focus on major combat operations (MCO) allows training to focus on general force preparation rather than specific force generation.[26] This trend has changed in recent years, however, with the focus at the CTCs shifting to preparation for operations in Afghanistan and Iraq.

---

[26] Braddock and Chatham, *Report of the Defense Science Board Task Force on Training Superiority and Training Surprise*, p. 21.

An Increasing Service Role in Force Preparation

In practice, training seems to be shifting toward a force-generation model as a result of the pressure of current military commitments and structural changes within the U.S. military. The onset of the war on terror amplified and exacerbated existing trends. An accelerating operational tempo caused the U.S. Air Force, and, later, the U.S. Army, to adopt the modular-forces paradigm and a rotational-readiness concept that resembled the readiness system of the U.S. Navy and the USMC. These developments required a degree of centralized management that undermined the practice of assigning operational forces to COCOMs—a practice on which the JTS appears to have been based. The requirement for centralized management was formalized by the DoD's adoption of the Global Force Management construct, in which JFCOM recommends sourcing solutions for operational requirements regardless of units' current assignment status. Finally, the demands of the wars in Afghanistan and Iraq require that engaged forces have to be relieved periodically, further undermining the JTS's presumption of permanency of assignment. The emerging reality is that the services provide the greatest degree of continuity and authority for operating-force units and, therefore, that they have inherited the de facto responsibility for planning and organizing training for their forces.[27]

None of this is to say that the services are disregarding the needs of COCOMs. Rather, the COCOMs simply have less practical ability to fulfill these with their own resources. Instead, the onus is on the services to recognize the requirements and work proactively to meet them.

---

[27] In the U.S. Air Force, the adoption of rotational readiness and the modular-forces concept occurred with the adoption of the air expeditionary force construct in the late 1990s. Air Combat Command manages air expeditionary forces and allocates them against regional COCOM requirements. In the U.S. Army, the transformation to a modular-force construct and the adoption of rotational readiness more or less coincided with the DoD's adoption of Global Force Management. The assignment of U.S. forces in the continental United States to JFCOM actually means that the services exercise de facto responsibility for their readiness and training. For example, the U.S. Army Forces Command commands almost 85 percent of the active U.S. Army, at least while the units are not in Iraq, Afghanistan, or elsewhere.

Methodology

The hypothesis for this study was that insights from other national military training and readiness models could inform improvements in U.S. training systems. To this end, RAND looked outside of the U.S. military training structure to identify insights from the military establishments of China, France, the UK, India, and Israel. Our purpose was to assess their training models and practices for meeting demands across the range of military operations. Our methodology for each nation included

- identifying its defense establishments
- identifying both its strategic demands and its strategic focus (including deployments)
- determining how the selected defense establishments prioritize demands (through, for example, examining their mission sets or the focus of their range of operations
- comparing component roles with mission sets
- assessing training programs for adapting forces
- determining unique training regimes for identified missions and assessing their utility for the U.S. systems.

This broad methodology resulted in several more-discrete questions in four categories. We answered each of these questions for each country, when possible:

- strategic imperatives
  - What are the strategic imperatives or demands, and which are given priority?
  - Are deployments (in and out of the country) a component of addressing these imperatives?
  - Which imperatives are similar to those of the United States?
- deployment training
  - How is deployment training addressed?
  - How do different deployment requirements affect doctrine, organization, training, materiel, leader development and education, personnel, and facilities (DOTMLPF)?

- How serious is the country about jointness and joint training?
- How does the country think about training for coalition operations?
- Are any programs similar to those used by the U.S. military?
- human resources
  - Is the force conscript, volunteer, or mixed?
  - Are there statutory limits on deployments (e.g., rules of engagement [ROE], conscript deployment status)?
  - How do recruiting and retention affect deployments?
  - How are forces manned (i.e., through unit or individual rotations)?
- specialty forces
  - Are there also purpose-designed or specialty forces, or are there only general-purpose forces?
  - How are specialty forces treated?
  - What roles do specialty forces play in enhancing training?
  - What are the strengths and shortcomings of specialty forces?

Again, the purpose of these questions was to identify the similarities and differences between U.S. and foreign models to gain insights into potential improvements to the U.S. system.

In our case studies we relied on open-source materials and unclassified interviews.

Monograph Structure

Chapters Two through Six contain the case assessments for China, France, the UK, India, and Israel, respectively. Chapter Seven provides a cross-case assessment, identifies insights from the foreign case studies, highlights areas that we believe require further analysis, and offers several recommendations for OSD.

CHAPTER TWO

China

Introduction

China's military has been engaged in a rapid transformation over the past decade. The emphasis of China's military planning has shifted from low-tech "people's war" to "informationized" local wars. The official Chinese defense budget quadrupled in real terms between 1996 and 2007; the size of the military decreased by nearly a quarter; and the equipment the military employs, which used to consist of 1950s-era Soviet designs, now includes modern, domestically produced systems.

The country's military training is being transformed as well. Chinese military leaders and analysts have been closely studying the training methods of other countries and attempting to implement what they perceive as the best practices that are appropriate to China's military. Although significant shortcomings in Chinese military training remain, a concerted effort to improve training practices is under way and is likely to continue in the coming decade.

The Defense Establishment

China's military is generally referred to in English as the *People's Liberation Army* (PLA). A more-accurate title, however, might be *People's Liberation Military*, as the PLA includes China's navy, air force, and strategic rocket forces, and the Chinese name, 中国人民解放军, refers to military forces in general, not just the Army [陆军].

The PLA is under the command of China's Central Military Commission (CMC), which is usually chaired by China's most senior civilian leader. Currently, it is chaired by the president of China and the secretary-general of the Chinese Communist Party, Hu Jintao. The other membership of the CMC is not fixed but currently consists of ten senior military officers.[1]

Below the CMC, the chain of command divides into several threads. First, there are four general departments: the General Staff Department, the General Political Department, the General Logistics Department, and the General Armaments Department. These general departments do not have direct line authority over any combat units, but they are important for setting overall policy, strategy, and regulations for China's military. They also function as the general HQ for the PLA Army, which, unlike the other services in the PLA, does not have a separate HQ.

Second, there are seven military region (MR) commands that represent the seven geographic regions of China shown in Figure 2.1. All PLA Army forces are directly under the control of one of these MR commands. Note that these are not warfighting HQ. In the event of a conflict, a "war zone" (theater) [战区] HQ would likely be established that incorporated ground, naval, air, and conventional surface-to-surface missile forces from more than one MR under a single unified commander. Finally, also directly under the CMC are HQ commands for each of China's three nonarmy services—the PLAN, the PLAAF, and the Second Artillery Force (China's strategic rocket force, which is technically an "independent branch," not a full-fledged service). This command structure is illustrated in Figure 2.2.

---

[1] As of late 2009, the CMC had two vice-chairmen—Guo Boxiong and Xu Caihou—and the other members of the CMC were Liang Guanglie (the minister of defense), Chen Bingde (the director of the General Staff Department), Li Jinai (the director of the General Political Department), Liao Xilong (the director of the General Logistics Department), Chang Wanquan (the director of the General Armaments Department), Jing Zhiyuan (the commander of the Second Artillery), Wu Shengli (the commander of the PLA Navy [PLAN]), and Xu Qiliang (the commander of the PLA Air Force [PLAAF]). See Government of China, "中华人民共和国中央军事委员会 [Central Military Commission of the People's Republic of China]," Web page, March 15, 2008.

Figure 2.1
Chinese Military Regions

SOURCE: U.S. Department of Defense, Office of the
Secretary of Defense, *Annual Report to Congress:
Military Power of the People's Republic of China 2008*,
undated [c. 2008].
RAND *MG836-2.1*

The Components

More than two-thirds (1.6 million) of the personnel in the PLA belong
to the PLA Army.[2] As previously noted, there is no separate HQ for
the PLA Army. Instead, all PLA Army forces are under an MR command. Depending on their type and function, however, PLA Army
units also come under the authority of one of the general departments.
The General Staff Department has authority over combat units and

_____

[2] China.org.cn, "[China's National Defense in 2006:] IV. The People's Liberation Army,"
Web page, undated; International Institute for Strategic Studies, *The Military Balance 2008*,
London: Routledge, 2008, p. 376.

Figure 2.2
PLA Command Structure

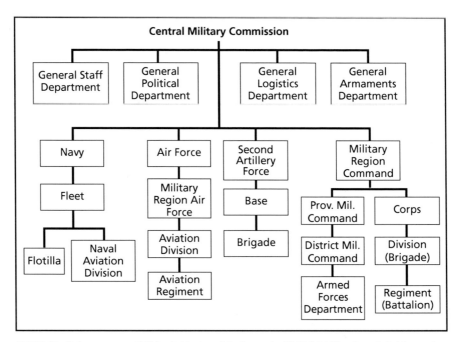

SOURCE: China.org.cn, "[China's National Defense in 2006:] IV. The People's Liberation Army," Web page, undated.
RAND *MG836-2.2*

their commanders,[3] the General Logistics Department has authority over logistics personnel and units, the General Armaments Department has authority over personnel and units responsible for managing weapon systems, and the General Political Department has authority over the PLA's political commissars.

Although most of the reductions in PLA force size in recent years have occurred in the PLA Army, and although the PLA Army is undergoing mechanization, it remains largely a light-infantry force. It consists of approximately 200 combat brigades (about four times as many as are in the U.S. Army), but it operates about 7,700 main battle tanks

---

[3]   National Air and Space Intelligence Center, *China: Connecting the Dots—Strategic Challenges Posed by a Re-Emergent Power*, Wright-Patterson AFB, Ohio: U.S. Air Force, 2007, p. 33.

(roughly the same number as the U.S. Army) and only about 5,000 armored personnel carriers and infantry fighting vehicles (roughly a fifth as many as the U.S. Army operates). The PLA Army has a huge artillery force with nearly 18,000 guns, most of which are towed, and multiple rocket launchers (about four times as many as the U.S. Army operates). The remainder of the PLA Army, however, consists mostly of foot soldiers or motorized infantry. The majority of the weapon systems operated by the PLA Army are still based on obsolete designs, but some (e.g., the Type 98/99 main battle tanks) are comparable in capability to the most-advanced systems used by the militaries of other countries.[4]

The PLAN has about 290,000 personnel and is divided into three fleets that generally operate in areas close to mainland China: the North Sea Fleet, the East Sea Fleet, and the South Sea Fleet. These fleets are under the direct command of PLAN HQ in Beijing (see Figure 2.2), but the fleet commanders are also deputy commanders of the MRs—the Jinan, Nanjing, and Guangzhou MRs—in which they are located and, thus, also report to the commanders of those MRs (who are always PLA Army officers).[5] It is likely that, as in U.S. unified commands, the fleet commanders report to PLAN HQ for activities associated with organizing, training, and equipping their forces and report to the MR HQ when participating in joint operations (such as joint exercises) within the MR. As previously noted, in the event of combat operations, a unified war-zone command would likely be created, and its assigned naval forces would report to the war-zone commander only, not to PLAN HQ.

The PLAN operates approximately 60 submarines (most of which are conventionally powered), 30 destroyers, 45 frigates, and a large number of smaller combatants. The PLAN has no operational aircraft carriers but maintains about 800 shore-based naval aircraft. As is the case with the PLA Army, the majority of platforms operated by the PLAN are outdated, but a few are modern systems comparable in capability to those operated by the U.S. military.[6]

---

[4] International Institute for Strategic Studies, *The Military Balance 2008*, pp. 30, 376–377; "NORINCO Type 98/Type 99 MBT," *Jane's Armour and Artillery*, February 28, 2008.

[5] Office of Naval Intelligence, *China's Navy 2007*, Washington, D.C., 2007, p. 58.

[6] International Institute for Strategic Studies, *The Military Balance 2008*, pp. 377–379.

The PLAAF has about 300,000 personnel and is divided between seven MR Air Forces (MRAFs) whose areas of operation correspond to the seven PLA Army MRs. MRAFs are under the direct command of PLAAF HQ in Beijing (see Figure 2.2), but the MRAF commanders are also deputy commanders of the MRs in which they are located and, thus, also report to the commanders of those MRs. As is the case with the PLAN, it is likely that the MRAF commanders report to PLAAF HQ for activities associated with organizing, training, and equipping their forces and report to the MR HQ when participating in joint operations within the MR. In the event of combat operations, a unified war-zone command would likely be created, and its assigned air forces would report to the war-zone commander only.

The PLAAF operates approximately 1,800 combat aircraft, fewer than 20 strategic-transport aircraft, fewer than 100 theater airlift aircraft, and 10 aerial-refueling aircraft. As is the case with the PLA Army and the PLAN, the majority of these platforms are outdated, but a few are modern systems comparable in capability to those operated by the U.S. military. Unlike the U.S. Air Force, the PLAAF also operates long-range surface-to-air missiles and the PLA's airborne corps of three divisions.[7]

The Second Artillery Force operates China's land-based nuclear missiles and, since the 1990s, has also had a conventional surface-to-surface–missile capability. The Second Artillery has about 100,000 personnel and, like the PLAN and the PLAAF, has its own HQ that reports directly to the CMC. However, the Second Artillery is not regarded as a full service on par with the PLAN and the PLAAF. Moreover, unlike commanders in the PLAN and the PLAAF, the commanders of the Second Artillery's subordinate units (called *bases* or *missile armies*—see Figure 2.2) are not also deputy commanders of the MRs in which they are located; rather, they report to the CMC solely through the Second Artillery's chain of command. This centralized command system is consistent with the government's desire to maintain tight control over China's nuclear forces, and China's nuclear missile forces would likely remain under the direct control of the CMC during wartime. It seems

---

7   International Institute for Strategic Studies, *The Military Balance 2008*, pp. 376, 380.

plausible, however, that conventional missile units would be put under the control of the war-zone commander during a conflict.

According to the DoD's 2009 annual report to Congress on Chinese military power, the Second Artillery possesses approximately 1,100 conventional short-range ballistic missiles (some of China's total stock of these weapons belong to the PLA Army, not the Second Artillery) and at least 150 ground-launched land-attack cruise missiles.[8] The newer models of these short-range ballistic missiles are highly accurate and provide China with a unique capability. Although the United States and, presumably, Russia have long had the technical ability to manufacture missiles with similar performance parameters, the United States and Russia are barred by their 1987 Intermediate Nuclear Forces Treaty from deploying ground-launched missiles with ranges of 500–5,500 km.

Human Resources

The PLA has approximately 2.3 million active-duty personnel. Of these, an estimated 1.2 million–1.6 million are enlisted personnel. Of the enlisted personnel, about half (600,000–800,000) are in their first two-year term of service; in Western analyses of the Chinese military, these personnel are often referred to as *conscripts*. In general, however, they are not true conscripts because their service is not involuntary. Most or all of them are, in fact, volunteers. To be inducted into China's military as a conscript today, rural residents must at least have graduated from middle school and urban residents must have graduated from a vocational high school [中专] or a three-year technical college [大专] or be enrolled in a four-year college [大学]. In China, approximately 11 million males turn 18 each year, and China therefore needs only about 3 percent of them to enlist in the military to provide the 300,000–400,000 recruits needed each year. (Note that women make up a relatively small portion of military personnel in China.) Middle-school education is now compulsory in China, and because military service is a relatively desirable career for rural residents in China, more

---

[8] U.S. Department of Defense, Office of the Secretary of Defense, *Annual Report to Congress*, 2009, p. 66.

than enough volunteers are generally available to meet rural recruiting quotas. Roughly one-third of recruits are supposed to come from urban areas and, given the higher educational standards for urban conscripts and the other employment opportunities for graduates of vocational high schools and technical colleges in China's booming economy, the PLA experiences difficulty in meeting its urban-recruiting quotas. It appears, however, that this difficulty simply results in a somewhat greater-than-desired proportion of recruits from poor rural areas and not in the involuntary conscription of urban residents.[9]

The PLA did not formally create a noncommissioned-officer (NCO) corps until 1999. Prior to 1999, the maximum period of service for enlisted personnel in the PLA was 16 years (it is now 30 years), and many duties performed by NCOs in other militaries were performed by officers in the PLA. NCOs currently constitute about half of the PLA's enlisted force, and the PLA is continuing to increase both the total number of NCOs and the duties these NCOs are capable of performing.[10]

The vast majority of the NCOs in the PLA are conscripts who volunteer to extend their duty at the end of their second year of conscription and are selected to be NCOs. A small percentage, probably less than 5 percent, are graduates of universities or three-year technical colleges who are recruited to join the PLA as NCOs without first spending two years as conscripts.

In 2008, it was mandated that future NCOs must have at least a high-school education and a certificate of professional qualification, which can be a degree from a vocational high school or a technical college (but not, it seems, a degree from an ordinary high school). Because a degree from a vocational high school or technical college is the minimum qualification required of urban conscripts, roughly

[9] Dennis J. Blasko, "PLA Conscript and Noncommissioned Officer Individual Training," in Roy Kamphausen, Andrew Scobell, and Travis Tanner, eds., The "People" in the PLA: Recruitment, Training, and Education in China's Military, Carlisle, Pa.: Strategic Studies Institute, 2008, pp. 103–106; National Bureau of Statistics, Statistical Yearbook of China 2006, Beijing: China Statistics Press, 2006, p. 104; Office of Naval Intelligence, China's Navy 2007, pp. 73–77.

[10] Office of Naval Intelligence, China's Navy 2007, p. 73.

a third of the personnel selected to be NCOs presumably had already satisfied the certification requirement when they were conscripted.[11] Of the remaining two thirds of NCO selectees, approximately 16,000 a year are enrolled in formal two- or three-year educational programs at PLA academies. Of these, about 10,000 are enrolled in vocational high-school programs and about 6,000 in technical-college programs. The latter set of selectees presumably enters the PLA with high-school degrees from nonvocational high schools.[12]

Because the maximum term of an NCO in the PLA is 28 years (plus the two years spent as a conscript), sustaining an NCO corps of 600,000–800,000 probably requires the creation of approximately 40,000–50,000 NCOs per year.[13] If a third of these enter the PLA already holding vocational high-school or technical-college degrees and another 16,000 are enrolled in such programs upon selection as an NCO, then another 11,000–17,000 need to acquire their certification of professional qualification (and, in some cases, high-school degrees) through some other mechanism. Each year, some NCO candidates who do not already possess a certificate of professional qualification and are not enrolled in formal educational programs at PLA academies are apparently sent to civilian colleges, research institutes, or industrial enterprises, and others may acquire their certificates through distance-learning programs. Nonetheless, it is not clear that the PLA will be able to satisfy its mandate that all new NCOs acquire some form of certificate of professional qualification.[14]

---

[11] This conclusion is based on the assumption that the ratio between urban and rural NCO selectees is approximately the same as the ratio between urban and rural conscripts. Statistics showing the actual proportion of NCO selectees who are from urban areas or who already possess degrees from vocational high schools and technical colleges were not available.

[12] Blasko, "PLA Conscript and Noncommissioned Officer Individual Training," pp. 111–113.

[13] This estimate was calculated by assuming that the year-on-year attrition of NCOs is a constant and that only 50 percent of NCOs make it halfway through the maximum 28-year term. At this annual attrition rate (4.83 percent), 40,000 new NCOs a year would sustain a total NCO corps of 621,000, and 50,000 NCOs a year would sustain a total NCO corps of 776,000. A lower actual attrition rate would mean a smaller requirement for new NCOs each year; a higher attrition rate would mean a larger requirement.

[14] Blasko, "PLA Conscript and Noncommissioned Officer Individual Training," pp. 116–117.

In addition to these basic education and certification requirements associated with becoming an NCO, NCOs in the PLA receive further training and education at various points in their careers through a variety of means. Examples include training sessions of one to three months at the PLA's NCO schools, short courses at training bases run by units at the regimental level and higher, distance learning with PLA academies or civilian institutions of higher education, and on-the-job training. The goal is for NCOs to have one specialty but multiple capabilities [一专多能]. Moreover, senior NCOs (i.e., those who have reached the top two of six total grades) are required to have a degree from a three-year technical college.[15]

Roughly one-third of PLA personnel are officers, whereas about 15 percent of U.S. military personnel are officers. The PLA is attempting to reduce this proportion by increasing the size and capability of its NCO corps, but this process is still under way. Officers in the PLA are acquired through two main mechanisms. The first is the PLA's own military academies, of which there are approximately 30. These academies accept students coming directly out of (civilian) high schools, current enlisted personnel, and officers who never acquired college degrees. Approximately half of the PLA's new officers (perhaps 15,000 a year) currently come from its own academies.[16]

The other half of the PLA's new officers come primarily from China's civilian universities. Most of these (11,000 in 2006) participate in China's National Defense Student program, a Reserve Officers' Training Corps (ROTC)–like program that was initiated in 2000, but some university graduates are recruited directly into the PLA to become officers without having participated in the program. Prior to the establishment of the National Defense Student program, about 3,000 officers a year were recruited this way, but the number of direct recruits may have diminished since that time. Furthermore, some portions of the PLA have established "2 + 2" programs in which college students spend

---

[15] Blasko, "PLA Conscript and Noncommissioned Officer Individual Training," pp. 111–117; Office of Naval Intelligence, *China's Navy 2007*, pp. 58, 82–83.

[16] John Corbett, Edward O'Dowd, and David Chen, "Building the Fighting Strength: PLA Officer Accession, Education, Training, and Utilization," in Kamphausen, Scobell, and Tanner, *The "People" in the PLA*, pp. 141–144.

two years at a civilian college and then transfer to a military academy for another two years.

Participants in the National Defense Student program are assigned to nearby military units during the summer to learn by watching, and, after graduation, they receive up to three months of political-military training before being commissioned as officers. Students recruited directly into the PLA after college graduation without participating in the program are apparently given an additional year of military training before being commissioned.[17]

In the past, some PLA officers were directly promoted from the enlisted ranks without attending college or an officer academy. Few if any officers are now created through this mechanism, however. Instead, the preference is to either retain highly capable enlisted personnel as NCOs or enroll them in one of the PLA's officer academies. As previously noted, some of the officers previously promoted in this way are currently being enrolled in the officer academies so that they can receive a formal undergraduate education.[18]

In addition to requiring that all new officers have a bachelor's degree, the PLA also encourages its officers to pursue continuing education, including obtaining masters and doctoral degrees from military and civilian universities in China, going abroad to attend foreign war colleges, and attaining various forms of nondegree education. Examples of nondegree education are short-term training programs in military topics, night classes at local civilian universities, specialized training classes provided by civilian universities, and self-directed study in reading rooms and computer laboratories maintained by military units.[19]

Recruiting and Retention Considerations

PLA recruiting follows a regular annual cycle. In August or September of each year, the General Staff Department and the General Political

---

[17] Corbett, O'Dowd, and Chen, "Building the Fighting Strength," pp. 141–142, 147–153; Office of Naval Intelligence, *China's Navy 2007*, pp. 69–70.

[18] Corbett, O'Dowd, and Chen, "Building the Fighting Strength," pp. 142–145; Office of Naval Intelligence, *China's Navy 2007*, p. 69.

[19] Office of Naval Intelligence, *China's Navy 2007*, p. 71.

Department consult with operational units to determine how many new conscripts and NCOs will be needed in the coming year. Conscription quotas are then assigned to each MR, which assigns quotas to each subordinate provincial military command, which assigns quotas to each subordinate military district, which assigns quotas to each subordinate People's Armed Forces Department. The last of these are military organizations that are staffed by both active-duty military personnel and civilian employees of the local government.[20]

All men who will reach the age of 18 by the end of the calendar year must register for military service, and, in theory, they remain eligible for conscription until they reach the age of 22; in practice, however, most conscripts are 18 years old. Women are not required to register for military service but may do so if they have graduated from high school and are between the ages of 18 and 19.[21]

As previously noted, military service tends to be more attractive to poor rural residents than to urban residents. This results in a skewing of demographics within the PLA: The proportion of military personnel who are from rural areas is greater than the proportion of the nation's total population that is from rural areas, and enlisted personnel consequently have lower average levels of education than might be expected. On the other hand, because a college degree is now required to become an officer in the PLA (and a high-school diploma is a prerequisite for a college education), it is likely that the PLA's officer corps includes a greater percentage of people from urban areas than the nation's population as a whole. Military service is, however, less attractive to rural residents than it used to be because it is no longer an assured path to a secure, well-paying job in a state-owned or village-owned enterprise upon demobilization. Thus, even among rural residents, the PLA can no longer count on being able to recruit the most-capable and most-motivated young men and women.[22]

---

[20] Blasko, "PLA Conscript and Noncommissioned Officer Individual Training," pp. 103–104; Office of Naval Intelligence, *China's Navy 2007*, pp. 74–75.

[21] Blasko, "PLA Conscript and Noncommissioned Officer Individual Training," pp. 103–104; Office of Naval Intelligence, *China's Navy 2007*, p. 75.

[22] Office of Naval Intelligence, *China's Navy 2007*, pp. 75–77.

The men and women who are selected for induction are notified in early November, and, in early December, they travel to their assigned training units, where they undergo induction training. The training units are established by division-level and brigade-level units or independent regiments (including their counterparts in the PLAN, the PLAAF, and the Second Artillery). The instructors are NCOs and junior officers selected from the parent unit. Induction training consists of 40 training days over two months, during which 60 percent of instruction is devoted to basic knowledge and skills (such as military regulations, marching, saluting, and small-arms operation) and 40 percent is devoted to political training (which includes military discipline and law, ideology, PLA traditions, and unit history). In addition, the recruits perform physical training, which builds up to a 5-km run three times a week. Upon completion of induction training, the new soldiers are integrated into operational units or, in some cases, signed up for specialty training.[23]

Manning Strategies

To ensure that military units are socially and politically independent of the regions in which they are located, most conscripts and new officers are assigned to units outside of their province of residence.[24] Once assigned, however, personnel tend not to circulate geographically. Instead, they tend to spend their entire careers with their original units and, when they are promoted to higher-level units, the new units tend to be superordinate to the original unit. Thus, a battalion commander is likely to come from one of the companies under that battalion. This system undoubtedly fosters unit cohesion but likely also causes units to be insular and resistant to changes mandated by the PLA's central leadership. It may also limit the dissemination of knowledge and experience within the PLA.

---

[23] Blasko, "PLA Conscript and Noncommissioned Officer Individual Training," pp. 106–109; Office of Naval Intelligence, *China's Navy 2007*, pp. 77–79.

[24] Blasko, "PLA Conscript and Noncommissioned Officer Individual Training," p. 104; Office of Naval Intelligence, *China's Navy 2007*, p. 76.

When they enter a military academy as cadets or join the PLA after attending a civilian college, PLA officer candidates are assigned to one of four career tracks: command, logistics, equipment, or technology. After serving as platoon or company-grade officers, they may be selected for a fifth track: political officer. Once assigned to a career track, officers tend to remain in that track for the remainder of their careers.[25]

The PLA has a total of ten ranks for its officers. The translated names for each of these ranks are the same as their U.S. counterparts, except for the rank of O-7 (the equivalent of a U.S. brigadier general), which is referred to as *senior colonel* or *senior captain*; an O-8 is still a major general or a rear admiral. There is a minimum time-in-rank requirement of four years before promotion to the next rank, except in the case of promotion from second lieutenant to first lieutenant (from ensign to lieutenant junior grade in the PLAN), for which the requirement is only two years.[26]

The PLA practices a strict up-or-out retirement system, but retirement ages are tied not to rank or time in service but to *grade*, which is a measure of the level of the position the officer holds. For example, platoon leader is a Grade 15 position; deputy division commander is a Grade 8 position. Thus, if an officer never advances beyond Grade 12 (i.e., deputy battalion commander), his mandatory retirement age is 40 regardless of whether he is a major or was never promoted above captain. The minimum time-in-grade requirement for promotion to each higher grade is three years.[27]

The PLA began creating reserve units in 1983, and, in addition to its 2.3 million active-duty personnel, it now has 500,000 reserve personnel. During peacetime, reserves are subordinate to the provincial or district military command of the province or district in which they are located. (District commands are subordinate to the provincial commands, and the provincial commands are subordinate to the MR commands; see Figure 2.2.) During wartime, they would be assigned

---

[25] Office of Naval Intelligence, *China's Navy 2007*, pp. 61–62.

[26] Office of Naval Intelligence, *China's Navy 2007*, pp. 2, 63.

[27] Office of Naval Intelligence, *China's Navy 2007*, pp. 63–64.

to a combat unit or come under the direct control of the war-zone HQ. China also has a civilian militia that would be mobilized in wartime to provide various essential support functions. The "primary militia" has about 10 million personnel, and the "ordinary militia" has another 100 million. The reserves and primary militia are composed mainly of demobilized military personnel.[28]

Strategic Demands and Focus

China's strategic demands and focus are a function of both the specific security problems the country faces and its current level of military capabilities. China's military forces are still in the process of modernization, and China does not yet have the resources to support a military able to influence and respond to events throughout the world. At the same time, China currently faces several potential military challenges along its periphery, the most significant of which is the possible requirement to coerce or invade Taiwan. As a result, China's strategic focus at present is primarily on challenges along the country's periphery.

Strategic Imperatives and Priorities

As a nation that shares land borders with 14 other countries, has several unresolved territorial disputes in nearby sea areas, and has several internal separatist movements, China has multiple strategic imperatives and priorities for its military. The most important of these is Beijing's desire to reabsorb the self-governing island of Taiwan, which has been politically independent since 1949. Although China's leaders hope for eventual peaceful unification with Taiwan, in the interim they perceive a need to retain the capability to deter any attempt by Taipei to formalize its independence and they appear to desire a capability to someday force Taiwan to politically unify with the mainland. This is a significant military challenge because Taiwan is 90 miles from mainland China at its nearest point and has substantial military capabilities. Most importantly, however, the United States has indicated that, under

---

[28] Office of Naval Intelligence, *China's Navy 2007*, pp. 57–58.

certain circumstances, it would come to Taiwan's defense in a conflict over Taipei's independence from Beijing.

In recent years, China has succeeded in resolving all of its land-border disputes (except with India), but the potential for friction with its neighbors remains. On China's northeast border is North Korea, China's only treaty ally. Relations between the two have been tense in recent years because Beijing has joined Washington, Tokyo, and Seoul in pressuring Pyongyang to give up its nuclear weapons. Beijing's greatest fear with regard to North Korea is that conflict will erupt again on the Korean peninsula, and its second-greatest fear is that the Pyongyang regime will collapse, which would likely send millions of North Korean refugees into northeast China and compel China to make difficult choices about how to ensure the security of North Korea's nuclear weapons.

To the north is Russia, a country with which China almost went to war in 1969. Relations between Beijing and Moscow are now cordial (at least in public), and Russia has been a major supplier of weapons and military technology to China. However, underlying suspicions remain. To China's southwest is India, with which China fought a border war in 1962. New Delhi continues to claim territory—some of which is occupied by China, and some of which is occupied by India—that Beijing regards as Chinese, and India has been strengthening its ties with the United States in recent years. To the south is Vietnam, which China attacked in 1979 and later engaged in a border war during the 1980s. Although all of these borders are currently peaceful, the history of conflict causes Beijing to remain wary of these neighbors.

China's maritime disputes include a dispute over the Diaoyutai Islands, which are currently occupied by Japan; a dispute over the location of the dividing line between China and Japan's exclusive economic zones and, thus, the ownership of potential oil and gas deposits in the disputed area; and overlapping disputes with Vietnam, the Philippines, Malaysia, Brunei, and Indonesia over the ownership of the islands and waters of the South China Sea.

China's police forces—especially the People's Armed Police, a paramilitary force under the joint leadership of the CMC and China's civilian State Council—have primary responsibility for suppressing and defeating China's internal separatist movements, but Beijing also needs to prevent these insurgencies from receiving support or sanctuary from outside China's borders. This means that China's relations with several neighbors with which it has no territorial disputes or history of conflict, such as Kazakhstan, Kyrgyzstan, Tajikistan, and Mongolia, nonetheless could become militarized. The Islamist movements in Afghanistan and Pakistan are of concern to Beijing because of their actual or potential connections with Chinese separatist and Islamist movements.

With this panoply of ongoing or potential security challenges on the country's periphery, China's military needs a variety of different capabilities. It needs to be able to (1) at a minimum, militarily coerce or punish Taiwan and, at a maximum, invade and conquer the island, possibly in the face of U.S. intervention; (2) defend China's borders against armed incursion or infiltration; (3) defend China's maritime interests; (4) conduct stability operations in North Korea; and (5) defeat internal insurgencies. Additionally, aside from these specific challenges, as a rising regional and, someday, global power, China desires the military capability to defend and pursue its national interests throughout the region and the world.

The Role of Out-of-Country Deployments in National Strategy

Out-of-country deployments have not traditionally played a major role in China's national strategy. From Beijing's perspective, China's last out-of-country *combat* deployment was its incursion into Vietnam in 1979, nearly 30 years ago. (Because it claims the entire South China Sea as Chinese territorial waters, Beijing would most likely regard the naval combat with Vietnam over Fiery Cross Reef in the South China Sea in 1988 as having occurred within China's borders.) Since 2000, however, China has gradually increased the frequency of its noncombat out-of-country deployments, which can be grouped into three categories: United Nations (UN) peacekeeping missions, exercises, and naval port calls.

The People's Republic of China has participated in approximately 20 UN peacekeeping operations, more than half of which it joined after January 2000. Prior to 2000, China's participation consisted mostly of providing only police forces or observers. Since 2000, how-ever, China has deployed more than 9,000 military personnel to over a dozen different UN peacekeeping missions. About 2,000 Chinese military personnel remained deployed in such missions as of this writing, a figure that can be compared with the total of fewer than 1,500 military personnel deployed during the entire period between 1990 and 2000. Nonetheless, even China's current level of involvement in UN peace-keeping represents an extremely modest amount of overseas deploy-ment for a military with 2.3 million personnel. Moreover, we note that none of the deployments has entailed significant combat action by the deployed Chinese forces.[29]

After virtually never exercising with other militaries prior to 2003, China has since conducted an average of three or four exercises annually with other nations, and about half of these exercises have been conducted outside of China's borders and nearby waters. Many of these events have been maritime search-and-rescue exercises and the size of the forces involved and the length of the deployments have been modest. Chinese ships also appear to make an average of five or six foreign-port calls each year. Thus, out-of-country deployments for exercises and port calls also appear to be extremely modest for a mili-tary of the PLA's size.[30]

---

[29] Ministry of National Defense, "[China's National Defense in 2008:] Appendix III: China's Participation in UN Peacekeeping Operations," Web page, July 31, 2009.

[30] See China.org.cn, "[China's National Defense in 2002:] Appendix II: Major Military Exchanges with Other Countries in 2001–2002," Web page, undated; China.org.cn, "[China's National Defense in 2004:] Appendix III: Major Military Exchanges with Other Countries (2003–2004)," Web page, undated; China.org.cn, "[China's National Defense in 2004:] Appendix V: Joint Exercises with Foreign Armed Forces (2003–2004)," Web page, undated; China.org.cn, "[China's National Defense in 2006:] Appendix II: Major International Exchanges of the Chinese Military 2005–2006," Web page, undated; Ministry of National Defense, "[China's National Defense in 2006:] Appendix IV: Joint Exercises with Foreign Armed Forces (2005–2006)," Web page, January 15, 2007; Ministry of National Defense, "[China's National Defense in 2008]: Appendix I: Major International Exchanges of the Chinese Military (2007–2008)," Web page, July 31, 2009; Ministry of National Defense, "[China's National Defense in 2008]: Appendix II: Joint Exercises and Training with For-eign Armed Forces (2007–2008)," Web page, July 31, 2009; Ministry of National Defense, "[China's National Defense in 2008]: Appendix III: China's Participation in UN Peacekeep-ing Operations."

How China's Strategic Imperatives Compare with Those of the United States

China's strategic imperatives differ from those of the United States in a number of significant ways. Perhaps the most fundamental of these dissimilarities is reflected in the dramatic difference in the importance of out-of-country deployments to the national strategies of the United States and China. U.S. strategic imperatives entail controlling and responding to events and developments far abroad. China's principle strategic imperatives are determined by challenges originating on the country's periphery or within its borders. China's strategic imperatives also entail preventing events that could directly threaten China's territorial integrity or the security of people living in China. In contrast, the events of September 11, 2001, notwithstanding, the United States' strategic imperatives since the end of the Cold War have largely revolved around developments that affect, or could affect, U.S. national interests but do not directly threaten U.S. territorial integrity or the security of people living in the United States.

Priorities: The Mission Set and the Range-of-Operations Focus

The PLA has a fairly broad set of missions (though it is not as broad as that of the U.S. military) and a comparably broad set of operations that it prepares to conduct.

The Missions of the Components

Broadly speaking, the PLA can be said to have seven principal combat missions: deterring a nuclear or other regime-threatening attack against China, preparing to blockade or otherwise coerce Taiwan, preparing to invade and occupy Taiwan, defending China's land borders, protecting China's coastline from invasion or attack, defending China's maritime territorial claims, and, in extremis, defending the government against domestic uprisings (as it did by putting down the Red

Guards in 1969 and the democracy protestors in 1989).[31] Because it controls most of China's nuclear weapons, the Second Artillery has primary responsibility for deterring a nuclear or other regime-threatening attack against China. The PLAN would be the lead service in a blockade of Taiwan, and the PLAAF and Second Artillery would be the lead service in other types of coercion against Taiwan. An invasion and occupation of Taiwan would be a truly joint campaign in which the PLA Army, the PLAN, the PLAAF, and the Second Artillery would all play indispensable roles. The PLA Army has primary responsibility for defending China's land borders and is the service that is called on to defend the government against domestic uprisings. The PLAN has primary responsibility for protecting China's coastline against invasion or attack and for defending China's maritime territorial claims.

The Range of Operations as Reflected in Preparations and Focus

An authoritative Chinese military publication identifies 17 different types of PLA "campaigns" (what the U.S. military would call *operations*) divided into joint campaigns, army campaigns, navy campaigns, air force campaigns, and Second Artillery campaigns. The joint campaign types are

- joint blockade
- landing
- anti–air raid.

The army campaign types are

- mobile warfare
- mountain offensive
- positional offensive
- positional defense
- counterterrorist stability operations.

---

[31] The PLA has also always had a mission of providing domestic disaster relief and humanitarian assistance. This mission has taken on new importance since the massive earthquake in Sichuan in May 2008.

The navy campaign types are

- destruction of enemy surface naval formations
- interdiction of naval lines of communication
- offensives against coral islands and reefs
- protection of naval lines of communication
- defense of navy bases.

The air force campaign types are

- air offensive
- airborne
- air defense.[32]

Only one Second Artillery campaign type—conventional missile strikes—is identified in the 2006 edition of this publication, but the 2000 edition identified another type—nuclear missile counterstrikes—that undoubtedly still applies.[33]

Each of China's MRs and its associated fleet or MRAF emphasizes a specific subset of operations. The Nanjing and Guangzhou MRs are, unsurprisingly, focused on amphibious operations. The Chengdu and Lanzhou MRs are focused on high-altitude operations, which is consistent with the fact that their forces would most likely be used to defend China's borders with India and Central Asia. The Shenyang, Beijing, and Jinan MRs are focused on long-range mobility and rapid-reaction operations, which is consistent with the fact that their forces would most probably be used to respond to rapidly developing contingencies near their borders or to reinforce contingencies occurring in other MRs. PLA forces also tend to focus on training for operations in the geographic environments close to where they are located, including coastal areas, cities, deserts, mountains, jungles, and islands.[34]

---

[32] Zhang Yuliang [张玉良], ed.,《战役学》 [*Campaign Studies*], Beijing: National Defense University Press, 2006.

[33] Wang Houqin and Zhang Xingye [王厚卿 and 张兴业], eds.,《战役学》 [*Campaign Studies*], Beijing: National Defense University Press, 2000.

[34] Dennis J. Blasko, *The Chinese Army Today: Tradition and Transformation for the 21st Century*, London: Routledge, 2006, pp. 147, 151–164; Dennis J. Blasko, "PLA Ground Force

An Assessment of Component Roles Against Mission Sets

The primary missions of the PLA Army are to prepare to invade and occupy Taiwan, defend China's land borders, and, if necessary, defend China's government against a domestic uprising. The primary missions of the PLAN are to prepare to invade or blockade Taiwan, protect China's coastline, and defend China's maritime territorial claims. The primary missions of the PLAAF are to prepare to support an invasion or blockade of Taiwan, prepare to execute a coercive campaign against Taiwan, and support the defense of China's coastline and land borders. The primary missions of the Second Artillery Force are to deter a nuclear or other regime-threatening attack against China, prepare to support an invasion or blockade of Taiwan, prepare to execute a coercive campaign against Taiwan, and support the defense of China's coastline and land borders.

The PLA Army

The training cycle of PLA Army units is strongly influenced by the annual recruiting cycle, which results in the turnover of a quarter of the PLA's enlisted personnel each winter. With a quarter of their enlisted personnel being replaced and their key NCOs and junior officers being assigned to training recruits, operational units are likely largely dormant between December and February, except perhaps for conducting individual training and drills. Once induction training ends, units must work to integrate the new personnel into the unit and rebuild the skills of the unit's existing personnel. For the PLA Army, this means that units focus on individual and small-unit training during February and March. The scale and complexity of training increases over the following months, however, and, beginning in April, units may deploy

---

Modernization and Mission Diversification: Underway in All Military Regions," in Roy Kamphausen and Andrew Scobell, eds., *Right-Sizing the People's Liberation Army: Exploring the Contours of China's Military*, Carlisle, Pa.: U.S. Army War College, 2007, pp. 321–323; Bernard D. Cole, "Chinese Naval Exercises and Training, 2001–2005: Reporting and Analysis," paper presented at the "CAPS-RAND-CEIP International Conference on PLA Affairs," Taipei, 2005, p. 15; Kenneth Allen, "The PLA Air Force: 2006–2010," paper presented at the "CAPS-RAND-CEIP International Conference on PLA Affairs," Taipei, 2005, p. 13.

to larger training areas for between several days and several months to participate in exercises that may involve multiple PLA Army branches and even other services. Annual evaluations, often involving live-fire exercises, are conducted in October and November.[35]

Training for amphibious operations (i.e., training to invade and occupy Taiwan) is a major focus of PLA Army training, particularly in the Nanjing, Guangzhou, and Jinan MRs. Two PLA Army divisions—the 1st Motorized Infantry Division in the Nanjing MR and the 124th Infantry Division in the Guangzhou MR—have been transformed into amphibious divisions and appear to be the units that would have primary responsibility for an amphibious invasion of Taiwan. Another 20 combat divisions and brigades (i.e., roughly a quarter of the PLA Army's maneuver units) appear to have received at least some degree of amphibious training. Some of this training occurs at or near garrisons, but China has at least four joint amphibious training areas along its southeastern coast to which units deploy for up to several weeks.[36]

Other types of training conducted by PLA Army units, particularly in China's western regions, include training for border-defense operations against overland incursions and for counterterrorism operations. For counterterrorism operations, dozens of small, highly trained antiterrorist units with special weapons and tactics have reportedly been established in PLA (and police) units across the country, and special training courses on terrorist techniques and countermeasures have been provided to PLA commanders. Counterterrorist-training scenarios include hostage crises; hijackings; bomb-detection and bomb-disposal situations; and chemical-, biological-, and radiological-weapon events. Heavy weapons, such as tanks, armored personnel carriers, artillery, and helicopters, are often employed to eliminate terrorist strongholds in these exercises. Coordination with People's Armed Police, militia, and local police units is also a key feature of these exercises.[37]

---

[35] Blasko, *The Chinese Army Today*, pp. 145–146.

[36] Blasko, *The Chinese Army Today*, pp. 151–156; Blasko, "PLA Ground Force Modernization and Mission Diversification," pp. 322–325.

[37] Blasko, *The Chinese Army Today*, pp. 156–157; Blasko, "PLA Ground Force Modernization and Mission Diversification," p. 322.

Aside from repelling a conventional invasion or conducting cross-border counterinsurgency operations, the PLA Army's mission of defending China's land borders could theoretically also entail occupying part or all of a neighboring country, such as North Korea, that was experiencing domestic unrest or regime collapse. In general, there appears to be little reporting about explicit training for this type of operation, although the 2005 combined Chinese-Russian exercise called Peace Mission 2005 was ostensibly based on a scenario in which China and Russia were part of a UN-approved intervention in a third country whose cities and rural areas had been seized by insurgent forces with links to international terrorism. Similarly, although putting down domestic uprisings is a possible mission of the PLA, there appears to be no reporting about the PLA (as opposed to police units) training for this type of operation.[38]

As previously noted, beginning each April, PLA Army units periodically deploy to larger training areas. Individual divisions, brigades, and regiments have their own training areas and firing ranges. There are also larger regional training areas and live-fire ranges for armored and artillery units throughout the country, and each MR has its own combined-arms training base. These training areas and bases are equipped with varying levels of communication, monitoring, and mapping technologies. During an exercise, coordinators and umpires follow each company or battalion to control the exercise and render judgments about the results of the company or battalion's actions. At some of these areas, laser technology is used to determine the results of individual fires. Some training areas have created permanent opposition forces that are designed to simulate the tactics of certain foreign militaries, and these opposition forces are allowed to defeat the visiting unit.[39]

The PLAN

PLAN training follows an annual cycle similar to that of the PLA Army. From November to February, after the demobilization of many NCOs

---

[38] Blasko, *The Chinese Army Today*, pp. 157–159, 162–163; Blasko, "PLA Ground Force Modernization and Mission Diversification," pp. 325–331.

[39] Blasko, *The Chinese Army Today*, p. 149; Blasko, "PLA Ground Force Modernization and Mission Diversification," pp. 333–334.

and second-year conscripts who have not been promoted to NCOs, units focus on practice in basic subjects. Ship units generally conduct single-vessel training in port or in nearby waters but may occasionally form small groups for training. Live gunnery against land targets may also be practiced. From March to June, after new conscripts have been absorbed and NCOs have completed short-term technical training courses, ship units conduct group training, during which several ships of the same type work together. Surface ships may shoot at drones or practice against fighters of the PLAN Air Force (PLANAF). From July to October, ship units engage in task-force training, during which ships of different types work together. Occasional exercises of three to five days' duration are held during this period, and a final evaluation exercise is held in October.[40]

PLANAF unit training does not appear to follow as obvious an annual cycle as PLAN ship training. This is likely because PLANAF pilots are all officers, whose annual turnover rate is significantly lower than that of the PLAN's enlisted force. PLANAF pilots, all of whom currently operate land-based aircraft, apparently average about 125 flying hours a year. In the case of fighter- and attack-aircraft pilots, these 125 flying hours may entail 160 or 170 flights because the average training flight lasts about 45 minutes. However, some units are increasing the average time per flight without increasing the total number of flying hours (possibly to limit the number of flying hours on the aircraft engines, which must be overhauled or replaced much more frequently than modern Western engines), resulting in fewer flights per year.[41]

The PLAN also has two marine brigades. Marine training appears to follow an annual cycle similar to that of the PLA Army and the PLAN ship units. Limited unit training occurs from November to February after most second-year conscripts and some NCOs are demobilized and as new conscripts are being trained. Once induction training is complete and the new conscripts are absorbed into the operational units, training progresses over the following months from

---

[40] Office of Naval Intelligence, *China's Navy 2007*, pp. 43, 92–93, 95; Cole, "Chinese Naval Exercises and Training, 2001–2005," pp. 10–12.

[41] Office of Naval Intelligence, *China's Navy 2007*, pp. 48–50.

individual training to small-unit training to larger-unit training. Two comprehensive training evaluations are conducted each year, one in early June and one in late November. These evaluations involve live maneuver on shore and at sea, including live-fire operations on shore and beach-landing operations.[42]

The training and exercises conducted by the PLAN support all of the service's primary missions of preparing to invade or blockade Taiwan, protecting China's coastline, and defending China's maritime territorial claims. Amphibious landing operations are a major focus of PLAN training, as are antisurface, antisubmarine, and anti-air warfare, all of which would be required to protect a Chinese invasion force, and subsequent reinforcements and supplies, from attacks launched by U.S. and Taiwanese forces. Shore bombardment and mine countermeasures, which may also be required, are also part of PLA training. Training in all of these areas also supports the PLAN's mission of defending China's maritime territorial claims. Training in antisurface, antisubmarine, and anti-air warfare supports the PLAN's mission of protecting China's coastline, and training in antisurface and anti-air warfare supports the PLAN's mission of preparing for blockade operations against Taiwan.[43]

The PLAAF

PLAAF training supports all of the service's primary missions of preparing to support an invasion or blockade of Taiwan, preparing to execute a coercive campaign against Taiwan, and supporting the defense of China's coastline and land borders. These missions require the capability to attack airbases, ground forces, shore defenses, airfields, naval facilities, and other military targets, along with political and economic targets; to intercept aircraft attempting to fly in or out of a nearby country (as in a blockade of Taiwan); and to defend forces and facilities on the Chinese mainland from attack aircraft and missiles. (It is not clear whether the PLAAF would also be involved in providing air defense for naval forces, such as an amphibious invasion force,

---

[42] Office of Naval Intelligence, *China's Navy 2007*, p. 94.

[43] Cole, "Chinese Naval Exercises and Training, 2001–2005," pp. 3–17.

or whether the PLAAF has an anti–surface-ship capability.) Because the PLA's airborne forces belong to the PLAAF, the PLAAF would also be responsible for conducting airborne operations in support of an invasion of Taiwan or in defense of China's land borders. All of these requirements are reflected in PLAAF training. Defense against air and missile attack has long been the focus of the PLAAF, and, since the late 1990s, the PLAAF has begun regular training in overwater operations, which would be required in an invasion, blockade, or coercive campaign against Taiwan. Various types of air-to-ground operations are receiving increasing emphasis in the PLAAF, and, for the first time, the PLAAF appears to be developing tactics and techniques for providing close air support. Airborne exercises are conducted regularly, and their scale and complexity have increased over time.[44]

The Second Artillery

Relatively little is known in the open literature about training in China's Second Artillery Force, the most secretive of the PLA's services. Second Artillery training does, however, appear to follow an annual cycle similar to that of the PLA's other services, with single equipment-item and single-subject training occurring during the winter months. After February, training increases in scale and complexity until entire brigades are engaged in mobile, all-weather, day-to-night training. This training presumably includes both nuclear and conventional missile forces and thus ensures that the force is prepared for not only the mission of deterring a nuclear or other regime-threatening attack against China but also the missions of supporting an invasion, blockade, or coercive campaign against Taiwan and defending China's coastline and land borders.[45]

---

[44] Allen, "The PLA Air Force: 2006–2010," pp. 9–12; Blasko, "PLA Ground Force Modernization and Mission Diversification," pp. 319, 332; Blasko, *The Chinese Army Today*, pp. 160–162.

[45] Kenneth Allen and Mary Ann Kivlehan-Wise, "Implementing PLA Second Artillery Doctrinal Reforms," in James Mulvenon and David Finkelstein, eds., *China's Revolution in Doctrinal Affairs: Emerging Trends in the Operational Art of the People's Liberation Army*, Alexandria, Va.: CNA Corporation, 2005, pp. 196–197.

Training Regimes

Like the rest of the PLA, training regimes are in flux in China. Nonetheless, certain features, described in the following sections, are clearly identifiable.

Overall Training Methodology

As previously suggested, the Chinese military appears to take the approach of focusing units on specific missions, especially in the case of the PLA Army. In addition to the two divisions that have been earmarked for an amphibious invasion of Taiwan, for example, other PLA Army units appear to specialize in mountain warfare, desert warfare, jungle warfare, etc. The PLAN does not appear to be as mission-specialized as the PLA Army in the sense that forces from all three fleets are likely involved in preparing to invade or blockade Taiwan, protecting China's coastline, and defending China's maritime territorial claims. Each fleet does have a distinct operating area, however, and although each of the fleets may be involved in defending China's maritime territorial claims, the specific claims and adversaries against which these claims must be defended vary. Similarly, the PLAAF's MRAFs likely focus on the following contingencies associated with their MRs: Taiwan contingencies and coastal defense for the Nanjing, Guangzhou, and Jinan MRAFs and border defense for the Beijing, Shenyang, Lanzhou, and Chengdu MRAFs, although some forces in the latter four MRAFs may also be earmarked for a Taiwan contingency. Finally, Second Artillery units are probably largely focused on training for specific conflicts, particularly since the ranges and basing locations of their missiles mean that the missiles are optimized for a specific opponent and may even have preprogrammed targets.

The current emphases in training methodology in the PLA are realism, complexity, and jointness. In the past, training was conducted largely by small units of a single type (e.g., infantry, frigates, or fighter aircraft) operating independently under benign conditions of familiar terrain, daylight, good weather, and either no adversary or an adversary whose actions were scripted and known. Now, training routinely occurs on unfamiliar terrain, at night, in bad weather, under intense electronic-warfare and cyberwarfare conditions, and against unscripted opposing forces. It frequently involves combined-arms or joint operations of relatively large scale (i.e., up to division size in the PLA Army or its equivalent in the other services). Another aspect of increasing the realism of training is the emphasis now given to training in logistics and support functions, such as supply, refueling, and equipment repair and maintenance.[46]

Rigorous evaluation and critique are also now a central part of training in the PLA. At one ground-force training base, for example, a 40-page report is written after each exercise to assess the results, and only 10 percent of the report can focus on positive aspects; the remaining 90 percent must discuss issues and problems identified during the exercise. Standardized tests and evaluations of unit performance are now imposed, and units that perform poorly are required to undergo remedial training.[47]

A final noteworthy emphasis of PLA training is integrating information warfare into PLA training and exercises. In addition to traditional information operations, such as electronic warfare, destruction of enemy communication and information systems, and computer-network operations, information warfare includes what the PLA calls the *three warfares*: psychological warfare, media warfare, and legal warfare. Psychological warfare largely entails traditional techniques, such as using propaganda leaflets and loudspeakers. Media warfare refers to public-information campaigns designed to mobilize China's populace, demoralize the enemy's populace, and educate Chinese military personnel on the importance of minimizing civilian casualties and other types of collateral damage. Legal warfare refers to efforts to argue that China's military actions are justified under international law and to

---

[46] Blasko, *The Chinese Army Today*, pp. 146–151, 164–167; Blasko, "PLA Ground Force Modernization and Mission Diversification," pp. 316–318, 331–335; Office of Naval Intelligence, *China's Navy 2007*, pp. 28, 37, 43–44, 49–50, 88–94, 100–101; National Air and Space Intelligence Center, *China: Connecting the Dots*, pp. 33–35; Allen, "The PLA Air Force: 2006–2010," pp. 10–11; Allen and Kivlehan-Wise, "Implementing PLA Second Artillery Doctrinal Reforms," p. 196.

[47] Blasko, *The Chinese Army Today*, p. 149; Blasko, "PLA Ground Force Modernization and Mission Diversification," pp. 317–318.

ensure that Chinese forces adhere to internationally recognized standards of conduct.[48]

Deployment Training

As previously noted, PLA units deploy abroad only infrequently, and, when they do so, it is either for exercises and port calls or for UN peacekeeping missions. The UN peacekeeping missions, moreover, generally involve noncombat forces, such as engineers and medical teams. Nonetheless, these peacekeeping forces apparently do undergo specialized predeployment training that lasts between three months and one year. For example, one article describes how, prior to deploying to Darfur, an engineering unit received three months of training in providing first aid, acclimatizing to the heat in Darfur, following UN regulations, and understanding Sudanese culture.[49] Another article describes how a Chinese transportation company spent nearly a year receiving "intensive training in shooting, field survival, land mine removal and first aid" prior to deploying to Liberia in 2002.[50] Interestingly, both of these units were reportedly specially staffed with soldiers who volunteered for the deployment, although how the soldiers were selected is not described.

Because the PLA has not engaged in combat operations for two decades, there is no recent record of how it trains in advance of planned combat deployments. The PLA does train for combat deployments, but because no combat operations are (presumably) currently planned, this training encompasses a range of contingencies that could arise rather than a specific planned operation. This does not mean, however, that PLA planning assumes that, in the event of combat operations, there would be a period of operation-specific predeployment training. Rather, because the PLA's theories of military conflict hold that crises

---

[48] Blasko, *The Chinese Army Today*, p. 164; Blasko, "PLA Ground Force Modernization and Mission Diversification," pp. 319–320.

[49] Jiao Xiaoyang, "Insight: Engineering Peace, Prosperity in Darfur," *China Daily* (online version), September 17, 2007.

[50] Xinhua News Agency, "Chinese Troops Ready for UN Peace Mission," November 19, 2003.

can develop rapidly and with little warning, the PLA assumes that its units would deploy for combat in their existing state of readiness at the time the crisis occurred. Because the conflicts that the PLA prepares for would occur on China's periphery, moreover, combat deployment generally means deployment *within* China. This type of deployment is in fact a major focus of PLA training. Mobilization and deployment, including of reserve and militia forces and on short notice, are a key emphasis of many ground-force exercises, with units frequently deploying by road, rail, and air across hundreds of kilometers to participate in an exercise. Similarly, surface-to-air-missile units and PLANAF aircraft now routinely practice deploying to and operating out of unfamiliar bases. Whether PLAN ship units and PLAAF aviation units also routinely practice short-notice mobilization and out-of-area deployments is not clear.[51]

Joint Structures and Training for Joint Operations

Training for joint operations is a consistent theme of PLA training. That said, the degree to which truly joint training is conducted appears to remain limited. In many cases, *joint* simply means forces from more than one service operating in the same area at the same time. (The Chinese term translated as *joint* is 联合, which literally means *united*.) Consequently, in 2004, the Chinese military coined a new term, *integrated joint operations* [一体化联合作战], to describe what the U.S. military would regard as truly joint operations: more than one service working together in a single operation. Integrated joint-operations training is still infrequent in the PLA, and significant gaps remain. For example, PLAN naval vessels and PLAAF aircraft never train together in providing joint air defense for surface ships. Indeed, it is not even clear whether the PLAN ships and PLAN aircraft train together for this purpose.[52]

---

[51] Blasko, *The Chinese Army Today,* pp. 147–149, 167; Blasko, "PLA Ground Force Modernization and Mission Diversification," pp. 321, 331; Allen, "The PLA Air Force: 2006–2010," pp. 12–13; Office of Naval Intelligence, *China's Navy 2007,* pp. 49–50, 89, 100.

[52] National Air and Space Intelligence Center, *China: Connecting the Dots,* pp. 33–35; Blasko, *China's Army Today,* p. 150; Blasko, "PLA Ground Force Modernization and Mis-

The PLA has a peacetime structure for joint operations in the form of the MR commands. As noted at the beginning of this chapter, MRAF commanders and fleet commanders are also deputy commanders of the MR in which they are headquartered, and the MR commander presumably has command over all joint operations, including joint training and exercises, in his MR. The PLA has also created several coordination zones in various parts of the country. These zones are large training areas (distinct from the combined-arms training bases and amphibious training areas mentioned earlier in this chapter) established for the specific purpose of allowing forces from more than one service to train together.[53]

As previously described, in the event of a conflict, a war-zone HQ would likely be established that incorporated ground, naval, air, and conventional surface-to-surface missile forces from more than one MR under a single unified commander. Whether there are permanent war-zone HQ facilities with permanent staffs, or whether members of the HQ staffs of different MRs, fleets, and MRAFs have been identified in advance as war-zone HQ staffs for particular contingencies and periodically exercise together, however, is not known.

Training for Coalition Operations
Other than the bilateral and multilateral exercises previously mentioned, we found no information on how the PLA trains—if at all—for coalition operations.

Training Methodologies for Foreign Militaries
Other than joint training exercises, the PLA carries out relatively little training of foreign militaries. In 2005, the PLA conducted a land-mine–clearance training program for the Royal Thai Army along the Thai-Cambodian border.[54] Foreign military officers also attend courses at China's National Defense University in Beijing, but these courses

---

sion Diversification," pp. 318–319, 339; Cole, "Chinese Naval Exercises and Training, 2001–2005," pp. 10–11; Office of Naval Intelligence, *China's Navy 2007*, p. 101.

[53] Blasko, *The Chinese Army Today*, p. 150.

[54] Ian Storey, "Thai Massage for China's Military Muscle," *Asia Times Online*, July 11, 2008.

focus on China's security strategy and policy, not on operational or tactical skills. We found no information on specific PLA methodologies for training foreign militaries.

Comparison with U.S. Regimes

The PLA's training regimes are, in many ways, similar to those of the United States. For example, although both militaries recognize the importance of joint training, in practice, both fall short of the ideal in this area, although the PLA undoubtedly falls farther short than the does the U.S. military. Both militaries also emphasize realism, complexity, and rigorous evaluation in their training but, again, the PLA undoubtedly falls well short of U.S. standards in this area. Finally, both militaries, or at least both armies, appear to employ a CTC model. This observation is particularly interesting in light of the claim by a group convened by the U.S. National Intelligence Officer for Conventional Military Issues, cited in Chapter One, that the high cost of establishing such training centers effectively denied U.S. adversaries the ability to improve their proficiency using this model. That China nonetheless employs CTCs suggests that either the capabilities of China's centers are inferior to those of the U.S. military or that the conclusions of groups convened by the National Intelligence Officer for Conventional Military Issues are inaccurate (or both).

Perhaps the most basic difference between the PLA and the U.S. military training regimes is captured by the force-preparation and force-generation models. As described in Chapter One, in principle, the U.S. military's training system follows the force-preparation model, but, in practice, it more closely resembles the force-generation model. The PLA's training system, however, actually appears to follow neither model, at least as we define them. That is, PLA forces train neither for some universal set of military operations nor for a single, specific operation. Rather, they appear to prepare for a particular *type* of operation or circumscribed range of contingencies.

For example, the PLA Army's two amphibious divisions, although obviously focused on an invasion of Taiwan, are probably trained for amphibious operations in general. They are probably not, however, well-trained for nonamphibious operations (e.g., inland maneuver warfare). Similarly, PLA Army units stationed in Tibet may be particularly focused on the possibility of another border war with India, but they are probably capable of high-plateau and mountain operations against any adversary.

PLAN units may be more versatile than PLAN Army units in that all of them probably train for the possibility of combat against the U.S. Navy, but PLAAF units are probably semispecialized in the same way that PLA Army units are. For example, units in the Nanjing, Guangzhou, and Jinan MRAFs are probably more proficient in over-water operations than units from the other MRAFs, and their training is probably tailored to the tactics and capabilities of the air forces of the United States and Taiwan, whereas training for units in the Chengdu MRAF probably focuses on Indian tactics and capabilities.

Although the abstract training models employed by the PLA and the U.S. military are fundamentally different, the practice is quite similar: The training of units in both forces falls somewhere between the extremes of training for a universal set of military operations and training for a single, specific operation.

Adaptability Training

Adaptability does not appear to be an explicit goal of PLA training. Indeed, as previously noted, the PLA appears to intentionally circumscribe the range of operations for which individual units train. Nonetheless, it can be argued that, within those limits, adaptability is an important *implicit* goal of PLA training. For inherent in the emphasis on realism and complexity in training is a recognition that the conditions under which actual combat operations are conducted are likely to be unpredictable and ever-changing. The goal of training on unfamiliar terrain, at night, in bad weather, under intense electronic-warfare and cyber-warfare conditions, against unscripted opposing forces, and with other services and branches is not simply to accustom military personnel to each of the specific conditions under which they might have to fight but also to increase their ability to fight under *any* set of conditions. Thus, although it cannot be argued that the primary goal of PLA training is adaptability in the sense that any PLA unit will be able to fight anywhere under any circumstances—clearly, PLA training is designed to maximize the capability of particular units to fight against specific adversaries in specific operating environments— nonetheless, adaptability is recognized as an important trait for PLA personnel and units.

Key Insights

The most general observation that emerges from this case study is that, although the U.S. military undoubtedly prepares to conduct a greater range of operations than any other military in the world, the training imperatives of other militaries are not simply a subset of those of the U.S. military. The United States enjoys unique strategic advantages that are widely recognized but whose implications for military training are easily overlooked. In particular, for most militaries, employment of their forces is not generally assumed to entail deployment abroad. Most militaries, including China's, focus primarily on conflicts that could occur on their immediate periphery. When they do deploy their forces abroad, it is usually for humanitarian operations, such as peacekeeping, or for small-scale, often low-intensity contingencies. The United States, in contrast, has not conducted an MCO in its own *hemisphere*, much less on its immediate periphery, for more than 100 years. At least since the end of the Spanish-American War, the United States has always fought in, and continues to prepare for, wars far from its shores.

Related to this is the fact that many of the wars the United States has participated in recently have, to some degree, been wars of choice. This has meant that the United States could often plan on being able to conduct a significant period of focused, precombat training prior to deploying its forces. China's military leaders do not appear to plan on having this luxury. Instead, they assume that their forces will fight in their existing state of readiness at the time a crisis erupts. For these reasons, not all training models that are effective for the U.S. military are applicable to other militaries, and other militaries may have little in the way of experience or practice to offer the U.S. military in such areas as predeployment training.

Within these parameters, what is most striking about the PLA is the degree to which its training regimes are similar to those of the U.S. military. This similarity exists, in part, because the PLA has deliberately studied and selected what its leaders believe are the best practices of the U.S. and other Western militaries. However, it is undoubtedly also partly due to the universal applicability of certain principles, such as the importance of realism and objective evaluation in training and the potential benefits of joint operations that are more closely integrated. Likewise, the features of the PLA's training regime that differ from those of the U.S. training regime are not necessarily the result of different thinking or philosophies but rather of different circumstances. The specialization of PLA units for different types of contingencies, for example, is probably a consequence both of the fact that China is faced with a finite number of contingencies—all of which would occur on China's immediate periphery—that would require the use of military forces and of the large size and limited mobility of China's military forces. Most forces in western China are unlikely to be involved in or needed for a Taiwan contingency and, therefore, can focus on the contingencies in which they would likely be involved. Thus, the fact that particular PLA units focus on particular contingencies may not have any significance for the U.S. military.

Finally, it is important to remember that the PLA has not been involved in even low-intensity combat operations for over 20 years, and the PLA of today is fundamentally different from the PLA of the 1980s. Thus, current PLA leadership has no recent combat experience, much less any experience in conducting combat operations while leading a modern military. They have, to a large extent, deliberately cast aside the doctrine and training models inherited from the Chinese Civil War and Korean War and are looking to the U.S. and other Western militaries for examples and guidance. That they are doing so can be taken as an indicator that the overall U.S. approach to training is viewed as a highly successful one; however, it also means that, at present, the PLA may have little to offer the U.S. military in terms of unique perspectives or insight with regard to training.

CHAPTER THREE

France

Introduction

Surprising as it may seem, the French view the world in much the same way as the Americans do. They see a similar international environment in which the greatest threats to France are indirect and result from failed states and instability. The French state that these conditions lead to transnational terror and crime, which can either affect France's vital interests or strike directly at France itself. France and the United States share similar objectives, including the development of a just and stable international order, protecting friends and allies, and maintaining influence on the world stage. This strategic outlook leads to an exceedingly similar range of military operations, albeit one in which stability operations and peace operations figure more prominently in French doctrine than in U.S. doctrine. The two countries share an almost identical view of the necessity of projecting national power, and France assumes that the venue for French action will likely be thousands of miles distant from French shores.[1]

---

[1] The key strategic documents for understanding French strategy and policy are the 1994 White Paper (Ministère de la Défense, "Livre Blanc 1994–2003," Web page, undated) and the 2003–2008 Defense Program (Sénat Français, "Projet de Loi, Adopté le 15 Janvier 2003, No. 49, Sénat, Session Ordinaire de 2002–2003, Projet de Loi Relatif à la Programmation Militaire pour les Années 2003 à 2008," January 2003). The former describes France's strategic outlook, France's objectives, and the conditions under which French armed forces are likely to be deployed. It has remained remarkably stable as a source of French policy over a period of considerable change. As in most democratic nations, funding for the armed forces in France lagged considerably behind aspirations, leading to a situation in which French forces were badly overstretched by the beginning of this decade. The 2003–2008 Defense

French training models and methodologies are very similar to those of the United States. Like the Americans, the French put their CTCs at the heart of their training. Like the United States, in France, simulation-based training for HQ staffs comprises a CTC in its own right. Also like the Americans, the French are moving away from a force-preparation model, which emphasizes training for operations in general, to a force-generation model, which, in France, orients training on the specific operations in which French forces are now engaged. The Commandement de la Force d'Action Terrestre (CFAT)[2] coordinates the activities of a number of French organizations in integrating lessons learned from operations into the preparation of forces for specific operational environments (including rotations at their CTC, the Centre de Préparation des Forces [CPF]), presenting an example of a thoroughly integrated system of force preparation. This systemic adaptation based on ongoing lessons learned complements a traditional French emphasis on individual and organizational adaptability, described later in this chapter.

Examining the French armed forces' operational performance, the most that can be said with any degree of confidence is that their method of generating operational capability works for them within their particular strategic context. Their experience in the highly complex and challenging civil war in the Côte d'Ivoire demonstrates the utility of their system. On the whole, they rely more on professional education and operational experience and less on collective training than does the U.S. military, especially when it comes to inculcating adaptability. They appear to be successful in inculcating adaptability in regimental (battalion) staffs and higher echelons. The French intervention in the Côte d'Ivoire (2002–present) provides a case in point. In coping with the operation's crisis in November 2004, French forces collectively responded very effectively, but there were several indica-

Program represented a bipartisan commitment to remedy this situation and to match the capabilities of French forces to the objectives set for them. Incidentally, the law also sets quite specific benchmarks in terms of training and readiness that the legislature checks annually. The 1994 White Paper is currently undergoing one of its periodic revisions, but major changes to its strategic logic and priorities do not appear to be in the offing.

2   CFAT is more or less equivalent to the U.S. Army's Forces Command.

tions that small units and individual soldiers coped somewhat less well. Moreover, a 2006 study by the Centre d'Études en Sciences Sociales de la Défense (C2SD) on the French armed forces' adaptation to their post–Cold War missions indicated that junior officers and enlisted soldiers are often uncertain about their missions and confused about their role in the operational environment.[3]

Nevertheless, we must note that most French operations take place within a fairly narrow strategic context. For the most part, French operations take place in Francophone Africa and other former French colonial possessions. French forces have operated in such places as the Côte d'Ivoire, Chad, and Lebanon for well over a century, limiting the degree to which French soldiers and the French military have to adapt to alien contexts. Second, although the complexity of their operations is second to none, such operations have not recently approached the intensity of U.S. operations in Afghanistan or Iraq. U.S. officials should remember these caveats when pondering the utility of French practices.

A Note on Sources

Resources for the study of French training are abundant, and, in some respects, exceed the extent of the already overflowing literature on U.S. training models and methodologies. First, French officials, both in the United States and in France, were extremely helpful in providing information and in identifying other sources for this research. Second, French military institutions make a considerable amount of information available online. In particular, we found *Doctrine*, the online journal of the French Army's Centre de Doctrine d'Emploi des Forces (CDEF), particularly helpful in illuminating both the current conduct of French training and the evolution of French military thought. The French Ministry of Defense's C2SD sponsors much in the way of interesting research into French military institutions and experiences. Fur-

---

[3] Vincent Porteret, Emmanuelle Prevot, and Katia Sorin, *Armée de Terre et Armée de l'Air en Opérations: L'Adaptation des Militaires aux Missions*, Centre d'Études en Sciences Sociales de la Défense, 2006, pp. 127–136, records the frustration and stress experienced by junior officers, NCOs, and other soldiers.

thermore, unfamiliarity with French poses only a very slight barrier to research because the French government makes considerable efforts to translate documents into English. In fact, the only area in which the researcher encounters difficulty is the subject of French Air Force doctrine, of which there is not much in the formal sense. Finally, the French legislature—both the Assemblée Nationale and the Sénat—take considerable interest in training, establishing training norms and standards, and closely monitoring the conduct and effectiveness of training. Legislative reports, testimony, and hearings are particularly rich sources.

Strategic Demands and Focus

Strategic Imperatives and Priorities

The official Web site of the French Ministry of Defense lists three principal strategic objectives:

- **Constructing a stable international order.** Defined in practical terms, this is supporting the development of the European Union (EU), maintaining ties with the North Atlantic Treaty Organization (NATO), and participating in the struggle against international terrorism.
- **Honoring French commitments.** These commitments either are bilateral or exist in the context of collective security. They include commitments to NATO, commitments to the European Security and Defense Policy, bilateral treaties, and implicit commitments to nations with historical and cultural ties to France.
- **Defending France's vital interests.** One such interest is the maintenance of French autonomy and freedom of action. More prosaically, the vital interests are defined as
  - preserving France's territorial integrity and the security of its maritime and aerial approaches
  - allowing the exercise of sovereignty and the protection of French citizens, in which category expatriates are expressly included

- maintaining the peace of Europe and the zones that border on Europe, including areas essential to economic activity, access to natural resources, and the ability to trade freely
- maintaining French power, an interest deemed essential to fulfilling France's implicit responsibilities as a UN Security Council member and as a nuclear power.[4]

The Role of Out-of-Country Deployments in National Strategy

Operational deployments play a key role in supporting these three strategic imperatives. French doctrinal publications note that although France faces no adversary directly on its borders, there are many sources of actual and potential crises that lie somewhat farther afield. Ethnic and religious tensions in weak and failing states threaten to disrupt the peace of Europe, and the existence of large and threatening arsenals, poorly controlled, could engender crises. Approaches to Europe (i.e., the Mediterranean basin, the Middle East, and Africa) are marked by chronic structural instability, and the ambitions of regional powers and the proliferation of weapons of mass destruction create risk. France also views international terrorism and transnational organized crime as threats. Together, these imperatives create a need to project power beyond French boundaries on a routine basis. Indeed, the need to project power was one of the key reasons that successive French administrations have continued a thoroughgoing program of defense transformation, including the shift from conscription to an entirely professional force. The French also recognize a strategic requirement to deploy forces under joint command and control (C2) in support of civil authorities.[5]

How France's Strategic Imperatives Compare with Those of the United States

The degree to which France's strategic outlook resembles that of the United States is remarkable. Both nations seek to promote a stable

---

[4] See Ministère de la Défense, "Les Objectifs Stratégiques de la France," Web page, undated.

[5] Ministère de la Défense, État-Major des Armées, Division Emploi, *Concept d'Emploi des Forces*, revised edition, October 8, 1997.

and just international order and capable institutions that reflect their values, especially respect for individual rights and the rule of law. The countries define their vital interests in a similar fashion, giving priority to the defense of the homeland but also including the preservation of access to vital raw materials and markets. Both view the maintenance of their relative power on the international scene as an independent objective, although the size of the United States makes preeminence a reasonable aspiration in a way that is simply not possible for France. Furthermore, France and the United States identify virtually the same primary threats to the international order and to their national interests: rogue states, transnational terrorism and criminality, the proliferation of weapons of mass destruction, and the many challenges posed by weak and failing states. Both see the Middle East as the primary geographic locus of tension. Most importantly, both reserve the right to intervene unilaterally and independently.[6]

The foregoing is not intended to deny that there are real and serious differences between U.S. and French strategy. These differences, however, reflect dissimilarities in perspective and application rather than outlook. France's greater proclaimed attachment to international institutions results from both its conviction that collective action is more effective in restoring stability and from the fact that France lacks the necessary resources to intervene unilaterally on the scale required to do so effectively. For example, Africa is far more important to France than to the United States because of both the French colonial heritage there and Africa's geographic proximity to France. Furthermore, although French officials (and French strategic planning) allow for the possibility that France will be involved in large-scale MCO, they assume that another country or entity will lead such efforts. Finally, they presume that irregular conflict will be the dominant mode of French military operations.[7]

---

[6] Compare the French strategic objectives found in Ministère de la Défense, "Les Objectifs Stratégiques de la France," with those identified in U.S. Department of Defense, *The National Defense Strategy of the United States of America*, Washington, D.C.: Department of Defense, 2005.

[7] For the French conviction that coalitions are more effective in resolving crises, see Ministère de la Défense, Armée de Terre, Centre de Doctrine d'Emploi des Forces, FT-01,

The French Ministry of Defense develops and maintains operational capabilities to perform four strategic functions:[8]

- **Protection.** Like the United States, France views the defense of the homeland as its highest priority. This includes defending France's air and maritime approaches and providing the full range of military assistance to civil authorities.
- **Prevention.** The French concept of prevention corresponds with the U.S. concept of engagement. French officials view proactive engagement to halt violence and instability, or to contain it at the lowest possible level, as the best guarantee of long-term peace. The French concept of prevention includes functions that the United States calls security cooperation and assistance.
- **Projection.** Most significantly, the French see the need to deploy joint task forces over thousands of miles on short notice and sustain them indefinitely as essential capabilities. Unlike the United States, France anticipates that such intervention will probably take place in a multinational context rather than with some multinational assistance. Again in contrast to the United States, the French assume that they will almost certainly not have to confront a near-peer competitor and will instead confront adversaries whose military capabilities are significantly weaker than France's.
- **Deterrence.** France maintains an independent nuclear deterrent that is directed primarily at state actors. In theory, French nuclear

---

Gagner la Bataille: Conduire à la Paix, January 2007, pp. 4–5. This publication seems equivalent to U.S. Department of the Army, FM 1, *The Army*, Washington, D.C.: Headquarters, Department of the Army, 2005, which describes the role of the U.S. Army. Like FM 1, FT-01 asserts the primacy of land forces. French officials have also acknowledged a shift in their emphasis from conventional combat operations to stability operations; see Rupert Pengelly, "French Army Transforms to Meet the Challenge of Multirole Future," *Jane's International Defence Review*, June 2006, pp. 44–53.

[8] This synthesis rests on a comparison between French "strategic functions" (found in Ministère de la Défense, "Les Objectifs Stratégiques de la France") and the description of how the United States accomplishes its objectives (found in U.S. Department of Defense, *The National Defense Strategy of the United States of America*). The principal difference between the two lies in the U.S. aspiration, not shared by France, to dissuade potential adversaries from developing competitive capabilities.

capabilities target the military, political, and economic centers of gravity from which an adversary derives its power rather than threatening the wholesale slaughter of populations. How nuclear weapons are supposed to achieve this degree of discrimination, however, is somewhat vague.[9]

The Defense Establishment

The Components

Human Resources. The transition to a professional force principally affected the French Army because the French Navy and the French Air Force were already effectively composed entirely of volunteers. The French decision to become an entirely professional force stemmed from the determination that the most likely and most important requirement for French forces would be service overseas.[10] From 1996 to 2003, the French Army went from a strength of approximately 266,000 soldiers and civilians, of whom about 25,000 could be deployed overseas, to a strength of about 166,000 soldiers and civilians, of whom approximately 100,000 could, in theory, be deployed. Moreover, volunteers in the French Army tend to stay in the service for an average of almost six years, increasing the force's overall level of experience. The French Army's objective is for each volunteer to serve eight years, and some "junior" enlisted soldiers accumulate well over a decade of experience.[11]

---

[9]   David S. Yost, "France's Evolving Nuclear Strategy," *Survival*, Vol. 47, No. 3, Autumn 2005 pp. 117–146. For the broader U.S. concept of deterrence, see U.S. Department of Defense, *The National Defense Strategy of the United States of America*, pp. 7–8.

[10]   Although the 1994 White Paper recognized a need for a much higher proportion of deployable units and soldiers, the decision to transition to an all-volunteer force did not follow directly from this requirement; rather, it represented a series of institutional compromises among various actors in French society and government. See Bastien Irondelle, "Civil Military Relations and the End of Conscription in France," *Security Studies*, Vol. 12, No. 3, Spring 2003, pp. 157–187.

[11]   For the increase in deployable end strength, see Ministère de la Défense, Armée de Terre, "Towards a Professional Army 2008," briefing, 2008. We deduced the extraordinary length of enlisted service from the fact that, in 2003, the French instituted a program to allow soldiers with 14–16 years of service to become NCOs without attending the École Natio-

The French NCO corps is of very high quality. The average NCO is about 35 years old, has graduated from an NCO academy, and has passed two fairly rigorous professional examinations to attain his or her current rank. In contrast to the U.S. Army, with its extensive NCO-education system, in the French Army, most professional development is the result of self-directed study and extensive operational experience. NCOs must pass the examination for the Brevet de Spécialiste de l'Armée de Terre to be promoted to sergeant and the examination for the Brevet Superior de Technician de l'Armée de Terre to be promoted to staff sergeant. French NCOs are also highly experienced. Most have been deployed more or less annually for the length of their careers. Finally, French NCOs tend to choose repetitive tours in the same unit, which tends to increase both the cohesion and overall level of experience of units.[12]

The French officer corps is similarly well educated and well seasoned. On the whole, French officers are more senior, both in age and rank, than their U.S. counterparts. The average French officer is about 39 years old, and the bulk of the officer corps is fairly evenly distributed between the ages of 33 and 53. Commanders at each echelon are about one grade senior to their U.S. counterparts; major generals (equivalent to U.S. brigadier generals) command brigades, and colonels command regiments (equivalent to U.S. combined-arms battalions). As in the U.S. Army, however, captains command companies.[13]

---

nale des Elèves Sous-Officiers d'Active; see André Thieblemont, Christophe Pajon, and Yves Racaud, *Le Métier de Sous-Officier dans l'Armée de Terre Aujourd'hui*, Centre d'Études en Sciences Sociales de la Défense, May 2004. We obtained the actual average length of service from M. Joël Hart, *Avis Présenté au Nom de la Commission de la Défense Nationale et des Forces Armées sur le Projet de Loi de Finances pour 2007 (No. 3341)*, Vol. IV, *Défense: Préparation et Emploi des Forces, Forces Terrestres*, 2006.

[12] See Ministère de la Défense, *Annuaire Statistique de la Défense—2006*, December 2006, sec. III.2.2. For the education system, see Thieblemont, Pajon, and Racaud, *Le Métier de Sous-Officier dans l'Armée de Terre Aujourd'hui*.

[13] See Ministère de la Défense, *Annuaire Statistique de la Défense—2006*, secs. III.2.2. and III.2.1. Assertions about rank structure are based on telephone discussions with U.S. Army officers who were liaisons to the French Army, August 2007.

Officer education in the French Army is both more rigorous and less widely available than in the U.S. Army. All officers attend one of the French military academies for three years; only a very small proportion is commissioned from other sources. Cadets enter the academies having already undergone two years of intense academic study at a preparatory school. The officers with whom we spoke characterized this period, conducted entirely under civilian auspices, as the single most challenging part of their precommissioning regimen. By the time they are commissioned, French officers have completed five years of postsecondary education and have earned the equivalent of a master's degree. After receiving their commission, they proceed to their branch schools, where they spend about a year. During that time, they receive a thorough grounding in all the technical and tactical aspects of their branch. Thus, by the time they report to their units, French Army officers have completed about six years of academic and military education; U.S. Army officers typically receive between four and a half and five years of such education.

French Army officers also attend a staff course, which is more or less equivalent to the U.S. Army's former Combined Arms Service Staff School. Thereafter, French officers' access to education is more restricted, with less than one-fifth of officers being able to attend the staff college. Such attendance, however, is a prerequisite for command at the regimental level and higher. Moreover, like NCOs, French officers must also pass two professional examinations to be promoted in due course throughout their careers.[14]

Throughout its system of professional military education, the Commandement de Formation d'Armée de Terre (CoFAT) makes extensive use of scenario-based training, another key contributor to the development of adaptability, according to the IDA study. Instructors present military problems to students, who then must rapidly develop solutions. The class or small group discusses the situation, and the instructor then changes the problem slightly and has the students react

---

[14] Telephone discussions with U.S. Army officers who were liaisons to the French Army, August 2007; École Militaire de Spécialisation de l'Outre-Mer et de l'Étranger, home page, undated.

to the altered circumstances. The scenarios used cover the full range of military operations.[15]

The principal difference between French and U.S. military education, especially for junior officers, is that there is simply more of it in the French system. French officers spend more time in an educational system and attain a higher level of academic qualification than their U.S. counterparts. They receive far more of their education from professional academics and draw on a wider range of sources. They receive at least another six months of technical and tactical training in their functional schools.

Statutory Considerations. Currently, the French armed forces face no statutory restrictions on the employment of French forces. Historically, the French government could not employ regular army units, manned by conscripts, outside of metropolitan France. One of the consequences of the 1994 White Paper, however, was that French authorities concluded that almost all significant operations would be conducted overseas. This conclusion was one of the primary reasons for the decision to professionalize French forces. Moreover, the president of the French Republic enjoys considerable authority to deploy these forces, informing the legislature as a matter of courtesy.[16]

Recruiting and Retention Considerations. Professionalization has not been without its pitfalls: The quality of recruits has declined over the past several years. A 2001 study identified a decline of approximately 5 percent in the number of recruits with the equivalent of a high-school diploma. French NCOs noted a similar decline in 2003, telling researchers that recruits were less physically fit, less motivated, and, in general, less well suited for military service than their predecessors. It is difficult to weigh the comprehensive impact of these countervailing trends, however.[17] It is also difficult to determine whether this

---

[15] Discussions with French Army officers, HQ, CoFAT, March 4, 2008.

[16] Olivier Camy, "Cours de Droit Constitutionnel Général," Web page, undated.

[17] For the decline in recruits' educational attainment, see Caroline Verstappen, "Sociologie: Effet des Évolutions Démographiques et Sociales," in Pierre Pascallon, ed., *Les Armées Françaises à l'Aube du XXIe Siècle*, Vol. III, *L'Armée de Terre,* Paris: Harmattan, 2003, p. 332. For NCOs' observations, see Thieblemont, Pajon, and Racaud, *Le Métier de Sous-Officier dans*

decline represents a temporary or long-term trend. On the one hand, the general level of educational attainment among French youth is projected to rise significantly over the next couple of decades. On the other hand, the attractiveness of a military career is also expected to decline, with fewer French citizens willing to disrupt family life with frequent moves and separations.

Manning Strategies. High levels of education, experience, and cohesion in the French armed forces may mitigate some of the intrinsic requirements for predeployment training. In general, French air and naval forces tend to be more experienced and tend to have served with greater continuity than their American counterparts. To reduce training costs, the French Navy and the French Air Force keep pilots in flying positions within the same units for a higher proportion of their careers than do their U.S. counterparts. French pilots typically spend at least seven years with the same squadron before being transferred.[18] Within the last decade, the French have successfully transitioned from a conscript-based to an all-volunteer force, vastly increasing the proportion and total number of troops available for overseas deployment.

Although the French do not practice lifecycle manning per se, their manning policies achieve many of the same effects. Enlisted service members serve their entire enlistment (up to eight years) in one regiment. When they reenlist, they reenlist for that regiment. In fact, the regiments recruit and train their own soldiers. Only some technical specialists attend centralized training at a French Army school. French NCOs tend to serve repeated tours in the same organization. Officers' tenure with a given regiment varies by branch. For instance, infantry officers remain in the same regiment until after they have commanded a company, and armor officers generally transfer after three years in the same regiment. Pilots spend about seven years in a single squadron. All these factors tend to reinforce unit cohesion.

---

l'Armée de Terre Aujourd'hui, pp. 62–65. One should probably accept the NCOs' observations with a degree of caution because sergeants' complaints about the decline of recruit quality and the relaxation of recruit training have always been a staple of NCO discourse.

[18] See John F. Schank, Harry J. Thie, Clifford M. Graf II, Joseph Beel, and Jerry M. Sollinger, *Finding the Right Balance: Simulator and Live Training for Navy Units*, Santa Monica, Calif.: RAND Corporation, MR-1441-NAVY, 2002.

French NCOs have a great deal of operational experience; many have been deployed annually over a period of 15–20 years. These cohesive units are led by officers who are senior to their U.S. counterparts not only in terms of rank but also in terms of age and experience. Although French soldiers have relatively fewer opportunities for professional military education, education is thorough, intense, and complemented by a system of professional examination to ensure technical and tactical proficiency.[19]

Priorities: The Mission Set and the Range-of-Operations Focus

The Missions of the Components

In statute, organization, and doctrine, the French armed forces are highly joint. The chief of staff of the armed forces presides over the joint staff and commands French forces in the conduct of operations through the Centre de Planification et de Conduite des Opérations (CPCO). He also commands the État Major Interarmées–de Force et d'Entraînement (EMIA-FE), which oversees joint training. The EMIA-FE also controls the Poste de Commandement de Force, a standing joint HQ focused at the operational level of war. Most professional military education, beginning with St. Cyr and culminating in the École de Guerre, is joint. In this system, the role of the components is to organize, train, and equip forces in their respective domains.

The French system, although joint, is clearly dominated by the French Army. The French Army's overall manpower levels and budget share are approximately equal to those of the French Navy and the French Air Force combined. Additionally, most armed forces chiefs of

---

[19] Our conclusions about unit manning and the relative seniority of French officers are based on telephone discussions with U.S. Army officers who were liaisons to the French Army, August 2007. For more on local recruitment, see Valerie Berrette and Benoit Saint Vincent, *Implantation Locale des Régiments: L'Expérience des Régiments Anciennement Professionnalisés*, Centre d'Études en Sciences Sociales de la Défense, September 2003, sec. 1.3.3.2. Our description of other French manning practices come from discussions with French officers, HQ, CoFAT, March 5, 2008.

staff come from the French Army, which also dominates most operations and provides most of the deployed forces and HQ. Furthermore, the French Army is responsible for preparing joint forces and individuals for operational deployments.

The Range of Operations as Reflected in Preparations and Focus

French and U.S. doctrines describe virtually the same range of military operations. France's PIA-00.200, *Doctrine Interarmées d'Emploi des Forces Armées en Operation*, describes the following three categories of operation (called strategic options), which are essentially identical to those described in the equivalent U.S. publication, Joint Publication (JP) 3-0, *Operations*:

- *Les actions de force* [forcible actions] correspond to the U.S. category major campaigns and operations and include combat operations to defeat a more or less conventionally armed and organized force.
- *La maîtrise de la violence* [mastery of violence] corresponds to the U.S. category *crisis response and limited contingency operations*. French authorities see the management of violence as the most likely employment of French forces.
- *Le soutien à la prevention et à la sécurité* [support to prevention and to security] corresponds approximately to the U.S. category military engagement, security cooperation, and deterrence.[20]

The primary difference between French and U.S. doctrine lies in emphasis. French doctrine describes the management of violence and the stabilization phases of an operation as the decisive aspects

---

[20] See Ministère de la Défense, PIA 00.200, *Doctrine Interarmées d'Emploi des Forces Armées en Opération*, 2003, Chapter 6. As in the United States, the relationship between joint and service doctrine in France is not always comfortable. Ministère de la Défense, Armée de Terre, TTA 901, *Forces Terrestres en Opérations*, 1999, recognizes four categories of operations: coercion, management of violence, direct defense of French interests, and homeland defense. Literally, however, the "ground defense," its component tasks, and associated operational functions, as described, correspond to homeland defense. As previously mentioned, the French view deterrence per se as a strictly nuclear function, but, in their view, deterrence does include actions intended to dissuade an adversary from taking aggressive action.

of operations. Another important difference lies in the much higher weight French doctrine accords to international law. For instance, PIA-00.200, *Doctrine Interarmées d'Emploi des Forces Armées en Operation*, includes long excerpts from the UN Charter and the North Atlantic Treaty as annexes to its chapter on the strategic context.[21]

There are also more-subtle differences between the French category *maîtrise de la violence* and the U.S. category *crisis response and limited contingency operations*. The French category implicitly assumes that the role of French forces is not to impose a French solution but rather to maintain a state of equilibrium between two or more contending parties to a conflict over an extended period of time to allow for political resolution. The parties in conflict may differ in nature, organization, and modes of exerting violence, and they can be governments, insurgent and terrorist organizations, or mass movements, to name a few. At an operational level, establishing and maintaining the appropriate state of equilibrium requires greater sophistication than simply dominating or destroying a belligerent's military capability, and doing so results in an expanded set of operational tasks: controlling physical space; controlling mass movements, which may involve military operations to strike at the source of unrest; and controlling armaments. Nevertheless, although managing violence clearly requires more of French commanders and their staffs, it paradoxically limits the scope of tasks required of French forces, whose primary requirements are to be able to dominate the ladder of escalation and demonstrate the conclusive ability to win any potential engagement, thereby preventing any belligerent faction from establishing a monopoly of violence.[22]

According to a 2006 study by the C2SD, the French Air Force has little in the way of written doctrine and perhaps even less regard for it. Indeed, to the extent that the French Air Force recognizes a need for doctrine, it considers the doctrine promulgated by NATO and the UK largely sufficient to meet its needs. Instead of doctrine, French air

---

[21] Compare U.S. Joint Chiefs of Staff, JP 3-0, *Joint Operations*, Washington, D.C.: Joint Staff, 2006, Chapter 1, sec. 5, with Ministère de la Défense, Armée de Terre, TTA 901, Chapter 4.

[22] Ministère de la Défense, PIA 00.200, paras. 06-46–06-49.

officers prize mental flexibility and the ability to improvise according to the needs of the particular situation in which they find themselves.[23]

Compared with U.S. doctrine and the demands it places on U.S. military leaders, French doctrine appears to demand a broader range of skills; however, it also requires French units to conduct a narrower range of tasks. An important caveat is that the lack of French doctrine for air operations may mean that relatively more effort is required to understand the nature of a given operation and its implications for the employment of air power. To manage violence, for example, French officers and HQ may require a fairly sophisticated understanding of the operational environment and its dynamics in order to assign tasks (such as the establishment of checkpoints or defensive positions) to subordinate elements; those tasks, however, may seem relatively elementary. Similarly, the legal complexities involved in maritime security may strain the captain of a vessel, but executing the resulting tactical tasks under benign conditions may seem relatively simple for his or her crew. On balance, French doctrine seems to imply the need to devote greater effort to leader development and C2 training than to collective unit training.

An Assessment of Component Roles Against Mission Sets

The French Army

As previously noted, the French Army dominates the country's military establishment and plays the principal role in protection, prevention, and projection. All land forces, including the Troupes de Marine, are part of the French Army.[24] A significant portion of the French Army is continuously deployed in a domestic-security role in the context

---

[23] Etienne de Durand and Bastien Irondelle, *Stratégie Aérienne Comparée: France, États-Unis, Royaume-Uni*, Centre d'Études en Sciences Sociales de la Défense, 2006.

[24] Troupes de Marine is more accurately translated as overseas soldiers rather than sea soldiers. Historically, regiments of the Troupes de Marine were eligible for overseas employment; units of the regular French Army were restricted to metropolitan France. Although current and former members of the Troupes de Marine often refer to themselves as the French Marine Corps, they are not, and have never been, under the jurisdiction of the French

of Opération Vigipirate, in which soldiers augment paramilitary and civilian law-enforcement and security personnel. French Army personnel and units also conduct the bulk of French engagement activities, such as training foreign forces.

The resemblance between U.S. Army doctrine and that of the French Army is particularly close. The former lists four major types of operation: offense, defense, stability, and support operations. The French also list four "tactical modes": offense, defense, security, and assistance. The principal difference lies in the fact that the French do not differentiate conceptually between assistance provided to domestic authorities and populations, which the U.S. Army calls support operations, and assistance to foreign governments and populations, which the U.S. Army calls stability operations. Moreover, as can be deduced from their name, French security operations are more narrowly focused on providing security and, thus, do not include reconstruction activities, which are included in U.S. stability operations. France's four tactical modes do, however, cover the same range of military activities included in the U.S. Army's operational doctrine.[25]

The French Navy

Conceptually, the French Navy plays a major role in supporting all four strategic functions. It supports protection through the concept and practice of its doctrine of maritime security, described in the next paragraph. The French Navy occupies a leading role in military diplomacy, conducting combined exercises around the world to improve France's relations with its key allies. It provides C2, air support, and transport capabilities to support projection; indeed, the French maritime component command based at Toulon organized and conducted Opération Baliste, the evacuation of noncombatants from Lebanon during the

---

Navy, and their primary function is not the conduct of land operations supporting a naval campaign.

[25] Compare Ministère de la Défense, Armée de Terre, TTA 901, Chapter 4, with U.S. Department of the Army, FM 3-0, *Operations*, Washington, D.C.: Headquarters, Department of the Army, 2008, Chapter 3.

summer and fall of 2006. Submarine-launched ballistic missiles are the other principal component of the French nuclear deterrent.

France's doctrine for naval operations appears to be very similar to that of the United States. The French, however, also include a concept of maritime security, which covers the enforcement of international norms and French national law on the high seas, including efforts to halt illegal immigration, narcotics trafficking, and maritime pollution. To support this concept, the French government has created the post of maritime prefect, a position that embodies and exercises French authority to act on the high seas. For the most part, maritime-security operations involve surveillance on the high seas integrated with enforcement operations in waters over which France has jurisdiction. Conceptually, however, the exercise and enforcement of maritime security is limited only by national and international law, not by geography.[26]

The French Air Force

The French Air Force plays a supporting role in protection, prevention, and projection and a principal role in deterrence. Maintaining the airborne component of the French nuclear deterrent, composed principally of air-launched cruise missiles, is perhaps the French Air Force's most-important and highest-priority mission. The French Air Force conducts many combined exercises with NATO and non-NATO partners; indeed, such exercises constitute a high proportion of its overall training activity. With regard to force projection, the French Air Force provides important capabilities for transport and limited combat support for deployed French Army forces. The French Air Force has given relatively little thought to its role in intelligence, surveillance, and reconnaissance in irregular warfare and therefore has little in the way of capability in those areas. Finally, the French Air Force is also responsible for controlling and securing the air approaches to sovereign French territory.[27]

---

[26] Ministère de la Défense, Marine Nationale, "Dossier d'Information Marine 2006: Les Missions de la Marine—La Sauvegarde Maritime," Web page, undated; Hervé Coutau-Bégarie, "La Sauvegarde Maritime: Réflexions sur un Nouveau Concept," *Revue Maritime*, No. 463, September 17, 2007.

[27] See Ministère de la Défense, Armée de l'Air, "Missions," Web page, undated. See also Stéphane Abrial, "A Highly Professional Air Force," *NATO's Nations and Partners for Peace*,

Specialty Forces: The Gendarmerie

The French have not, to date, created any specialty units on the order of military-assistance-and-advisory groups or stability-operations units, but their gendarmerie offers some unique capabilities in support of full-spectrum operations.[28] Organized under the Ministry of Defense, the gendarmerie's primary mission is the maintenance of public order and security in France and in French territories overseas. The national gendarmerie nonetheless recognizes three primary missions associated with contingency operations:

- provost (the enforcement of French laws on deployed French forces)
- the maintenance of public order in an area of operations during the course of an intervention
- assisting in the transition to civil authority (specifically, the development of civil law-enforcement functions) in a host country.

Although the gendarmerie recognizes these missions, it contains no organizations devoted to them. This indicates a requirement for a period of organization and predeployment training in the event that French authorities decide to employ the gendarmerie in this role. Moreover, although the gendarmerie is a component of the Ministry of Defense, the French defense program decided to shift responsibility for the day-to-day management and training of the gendarmerie from the Ministry of Defense to the Ministry of the Interior in 2003.[29]

Vol. 52, No. 2, 2007; Michel Forget, "Spécificité du Rôle et des Contraintes des Forces Aériennes," *Penser les Ailes Françaises*, No. 13, April 2007, pp. 34–41.

[28] French Army doctrine for stability operations states that the French do not envision creating any such specialized units. French Army leadership has instead decided to emphasize versatility (Ministère de la Défense, Armée de Terre, *Doctrine d'Emploi des Forces Terrestres en Stabilization*, November 2006, p. 23).

[29] Ministère de la Défense, Gendarmerie, "Missions de Maintien de la Paix," undated. See also Sénat Français, "Projet de Loi, Adopté le 15 Janvier 2003, No. 49."

Training Regimes

Overall Training Methodology

In general, the French armed forces employ a force-preparation model that is conceptually similar to the formal U.S. model and includes a fairly small, but growing, force-generation component. Unlike the United States, in France, the separate services retain command of their units until they are deployed on operations, at which point they come under the command of the joint-force commander. Doctrinally, preparation for operations is considered a service responsibility and is supposed to take place before deployment.[30] The services, therefore, separately retain responsibility for preparing their forces for joint operations, and the principal difference between the French and U.S. models is that policy, not contingency, is responsible for this outcome in France. The joint-force staff retains responsibility for training its joint-force HQ (JFHQ), which is the primary objective of most joint exercises. The separate services seem to employ entirely different models and methodologies for preparing their capabilities for operations, but they do participate in what the U.S. would consider multiservice training. For instance, although the French Army tends to rotate forces into and out of operations as units, the French Air Force tends to rotate forces as individuals.[31] Thus, just as is the case with the United States, understanding France's models and methodologies for preparing for full-spectrum operations requires understanding the component services' models and methodologies. However, the extent to which the French rely on units' and individuals' previous operational experiences to prepare them for future operations cannot be overstated.

Operational experience plays a key role in generating and sustaining individual and collective proficiency in military tasks. As is discussed in a later section, the French Army uses a 16-month readiness cycle during which a unit usually deploys overseas for four to six

---

[30] Ministère de la Défense, Armée de Terre, Centre de Doctrine d'Emploi des Forces, *Doctrine du Processes, Reception, Stationnement, Mouvement, Integration*, 2006, pp. 16–20.

[31] Porteret, Prevot, and Sorin, *Armée de Terre et Armée de l'Air en Opérations*, pp. 107, 116. This study principally assesses military attitudes to peacekeeping and stability-operations missions, but it does briefly address different patterns of deployment.

months. Usually, units deploy at least twice before experiencing significant personnel turnover. This means that units and individuals are frequently challenged to adapt to new missions, new organizations, and
new operational environments, which, in theory, should contribute to
the force's overall adaptability. The French Navy operates on a two-year
cycle, and, although the French Air Force does not employ cyclic readiness, aircraft crews can expect to deploy frequently. In this context,
training serves more to refresh skills acquired during operations than
to establish those skills in the first place.

To the extent that they exist at all, interagency preparations for
operations remain ad hoc. A 2007 study by the C2SD criticized a general absence of interagency coordination in French operations overseas
and noted that the different agencies of the French government lacked
standard practices, procedures, and mindsets for coordinating their
operations.[32] In this respect, at least, it seems that the United States is
somewhat ahead of France.

Operational deployments on peacekeeping, presence, and
stability-operations missions are a fact of life for French forces, but
training tends to emphasize traditional combat operations. This is
both a hedge against the unexpected and a means of allowing French
forces to dominate the ladder of escalation. Until the middle of 2007,
the French Army did little to prepare units systematically for specific
operational environments.[33] The French Army has increasingly centralized and consolidated collective training under the CPF. Although unit
commanders still, at least nominally, select their unit's training objectives, in practice, this seems to create a situation in which their CFAT
uses the CPF to prepare units for operations following a more-or-less
standardized template. Therefore, it is useful to assess the efficacy of
this system of centralized and consolidated training.

The attainment of training standards is an important goal of
the management of French training, and these norms are focused on

---

[32] Niagale Bagayoko, and Anne Kovacs, *La Gestion Interministérielle des Sorties de Conflits*,
Centre d'Études en Sciences Sociales de la Défense, 2007, p. 9.

[33] Jean Kergus, "La Prise en Compte des Spécificités de la Stabilisation dans les Exercices de
Préparation et d'Évaluation," *Doctrine*, No. 12, August 2007, p. 47.

the individual services. The 2003–2008 Defense Program establishes annual training goals of 100 days, exclusive of operations, for ground forces; 180 flight hours for fighter pilots; 200 hours for French Air Force helicopter pilots (180 for the French Army); and 400 flight hours for transport pilots. Although training norms address joint training, they do so almost exclusively in terms of C2 training. Moreover, the French seem to place relatively greater emphasis on multinational training than on joint training.[34]

Deployment Training

For all intents and purposes, deployment training in France is a service responsibility. Although the French armed forces conduct many joint and combined exercises, these exercises appear to be oriented on military diplomacy rather than on actual preparation for operations. In any case, joint exercises are certainly focused on a scale that vastly exceeds that of most operations French forces actually encounter. Thus, although this chapter describes joint training organizations and processes, the individual services conduct most preparations for operations.

Joint Structures and Training for Joint Operations

American armed forces are either assigned or apportioned to COCOMs on a continual basis. Forces assigned to a particular geographic COCOM orient their training on contingencies likely to occur within that commander's area of responsibility, and they conduct that training with other service forces with which they are likely to operate. In contrast, the French organization of joint task forces takes place only within the context of a specific mission. Obviously, the organization of joint forces takes into account the nature of the mission, the operational environment, and the capabilities of any coalition partners. At the direction of French political authorities, the armed forces chief of staff, in conjunction with the chiefs of the respective service staffs,

---

[34] M. Guy Tessier, "Rapport d'Information Dépose en Application de l'Article 145 du Règlement par la Commission de la Défense Nationale et des Forces Armées en Conclusion des Travaux d'une Mission d'Information Constituée le 29 Mars 2005, sur le Contrôle de l'Exécution des Crédits de la Défense pour l'Exercice 2005," No. 2985, March 29, 2006, p. 28.

develops planning guidance for an operation. This planning guidance specifies the desired end state, the forces and resources allocated to its accomplishment, other constraints and limitations, and, notably, the ROE.[35] The planning guidance also names the operational commander. If it is a multinational operation, this process involves extensive consultation with other governments and their armed forces. The armed forces chief of staff then issues this planning guidance to the commander of the CPCO. The CPCO, which is roughly equivalent to the UK's Permanent Joint HQ, takes this guidance and uses it to develop a campaign plan and refine the troop list. It also develops supporting movement and deployment plans. Based on this campaign plan, the operational commander develops his or her orders and begins execution of the operation.[36]

In other words, none of the forces or HQ that will conduct a given joint operation have been either organized or oriented on their mission before the initiation of that operation. Therefore, it is possible to identify forces that will deploy to an ongoing operation and orient them on their mission. The responsibility for doing so rests with the services. The operational-level HQ, which may not even deploy from metropolitan France, may be organized around the JFHQ (i.e., the force HQ [FHQ]), which is based at Creil, or it might be organized around a component HQ. For that matter, the operational-level HQ may be multinational or may even combine tactical and operation foci in one organization.[37] In any case, the HQ in question will have to

---

[35] The French armed forces accord particular importance to ROE, which Americans generally view as an operational and tactical constraint. That is not to say, however, that the French are reluctant to use force. The French have a maxim: "Either we fire or retire." That is, if force is not an option, the forces should not be there in the first place. See Shaun Gregory, "France and *Missions de Paix*," *RUSI Journal*, Vol. 145, No. 4, August 2000, p. 62.

[36] Giles Rouby, "The Joint Dimension of Operations Command," *Doctrine*, No. 5, December 2004. Le Centre de Doctrine d'Emploi des Forces publishes *Doctrine* as a military professional journal on various topics of interest. See also Jean-Pierre Teule, "Le CPCO au Cœur de Nos Opérations," *Revue Défense Nationale*, May 2007, pp. 64–69.

[37] For example, during Opération Artemis, the French deployment to Uganda in 2003, the FHQ provided the nucleus of an operational HQ that remained in Paris. Interestingly enough, the lingua franca of that HQ was English, the language in which the FHQ aspires to be able to operate on a routine basis. See Bruno Neveux, "Command and Control for

integrate individual augmentees and spend some time orienting on the mission. Further, although the French have a joint-exercise program, preparation for operations is a service responsibility, a situation that closely resembles arrangements in the U.S. armed forces.[38] In sum, each French joint task force is assembled for each specific operation out of what is available at the time, which would seem to complicate joint training and integration. Indeed, French joint doctrine recognizes the resulting probability of friction in establishing joint task forces as a "délai de mise sur pied [delay in establishment]."[39]

Several factors help compensate for the inevitable friction, however. The first is the relatively small size (less than 350,000 total personnel) of the French armed forces, which increases the chances that units and individuals will have previously trained or served together (or both). Second, the French strategic focus is relatively narrow, being concentrated on Eastern Europe and francophone countries in the Middle East and Africa.[40] Third, the French armed forces, especially the French Army, pay particular regard to what they call the principle of modularity, according to which forces are organized with the neces-

---

Operation Artemis," *Doctrine*, No. 5, December 2004. For information on the combination of tactical and operational levels of command, see Ministère de la Défense, État-Major des Armées, PIA 00.102, *Concept du Niveau Opératif,* July 2004, p. 7. For the organization of a component headquarters as a joint-task-force HQ, see Ministère de la Défense, Armée de Terre, TTA 901, sec. 1.2.1.2.

[38] France's air forces fall under the purview of the Commandement de la Défense Aérienne et des Opérations Aériennes, which combines the responsibilities of the U.S. Air Force's Air Combat Command and the Air Mobility Command. French naval forces, however, fall under four geographical commands. See Ministère de la Défense, Marine Nationale, "Le Commandement Opérationnel: La Conduite des Forces," Web page, undated.

[39] French doctrine acknowledges the existence of this friction, noting that high-stress, no-notice missions require a degree of cohesion not necessarily inherent in the course of the readiness and training center. Such missions require additional preparation, including joint training. See Ministère de la Défense, Armée de Terre, TTA 901, sec. 9.1.2.2.2, and Ministère de la Défense, PIA 00.200, paras. 03-39–03-42.

[40] French defense planning and programming are based on five scenarios that describe the circumstances under which French forces are likely to be engaged. With the exception of "actions in support of international law," all of the scenarios list these areas as the probable location of operations. See Ministère de la Défense, État-Major des Armées, Division d'Emploi, *Concept d'Emploi des Forces,* Chapter 4.

sary support and sustainment to execute their functions at the lowest level possible. (In the U.S. vernacular, this arrangement is known as plug and play.) Modularity, however, seems to principally concern the French Army.[41] Finally, French authorities remark that their forces will probably be conducting operations of lower intensity, which reduces the requirement for a high degree of cohesion prior to deployment.

In short, considerable potential for friction seems inherent in French procedures for organizing and deploying joint task forces. Understanding how and whether French predeployment training attempts to reduce that friction can, therefore, offer both insight into the training's utility and lessons for the United States on how to refine its own processes.

Most French joint training is of the joint-staff collective-training variety. French joint training falls under the purview of the EMIA-FE. With rare exceptions, joint training and exercises focus on force preparation in a multinational context. French statute prescribes a rigorous program of joint training, as indicated in Table 3.1.

Table 3.1
French Joint-Training Norms

Service	Principal Measures of Performance	Annual Norms to Be Attained by 2008
French Army	Joint and multinational exercises	16 exercises that involve either corps, division, or brigade HQ
French Navy	Joint and multinational exercises	1 major NATO exercise; one major EU exercise (every 2 years); 5 joint exercises
French Air Force	Joint and multinational exercises	1 major international exercise per pilot (every 2 years for transport pilots)

SOURCE: Sénat Français, "Projet de Loi, Adopté le 15 Janvier 2003, No. 49."

---

[41] On the other hand, the principle of modularity seems to be honored more in the breach than in the observance. In 2003, a French study concluded that units were deployed more often as platoons or even sections than as companies (let alone battalions, which are, supposedly, the smallest coherent module in the French Army). See Thieblemont, Pajon, and Racaud, *Le Métier de Sous-Officier dans l'Armée de Terre Aujourd'hui*, p. 35.

In the French Army, every brigade, division, and corps HQ is supposed to participate in one joint exercise annually; the French Navy is supposed to participate in a total of about six joint exercises annually. Interestingly, French law directs that French pilots participate in a joint exercise regularly but says nothing about French Air Force C2 assets. From one perspective, these norms reflect an assumption about the likelihood and frequency of joint operations in the land, maritime, and air domains, with joint and multinational land operations presumed to be quite likely but operations of a predominantly air component seen as barely possible. According to statute, the primary measure of an exercise's "jointness" is the integration of core elements from either the strategic HQ (i.e., the CPCO) or the operational HQ (i.e., the FHQ). Joint exercises, shown in Table 3.2, rarely involve either service HQ or actual maneuver forces.

Of the seven major joint exercises listed on the Web site of the État-Major des Armées and reproduced in Table 3.2, only two involved actual maneuver forces. Both of these were bilateral exercises with Persian Gulf states, Kuwait and the United Arab Emirates. All of the joint exercises listed on the Web site centered on C2 and employed the EMIA-FE's own assigned FHQ at Creil. Operationally, the scenarios tended to focus on crisis management or peace operations. Understandably, none of these exercises oriented on likely eventualities. Given the political sensitivity of such multinational exercises, French officials are loathe to introduce much uncertainty into the script; however, this renders the exercises less valuable as a training vehicle for adaptability. At present, the French armed forces neither have nor envision adopting anything analogous to the U.S. JNTC capability.[42]

French forces do conduct joint training exercises while deployed on operations, at least when those operations are of sufficiently low intensity. During the last couple of years, for instance, French forces in Côte d'Ivoire carried out several joint exercises to project a battalion

---

[42] Lefebvre, "Pourquoi les Armées Doivent-Elles se Doter d'un Centre d'Entrainement Tactique Permanent Interarmées et International en France?" *La Tribune du CID*, October 2006.

Table 3.2
Major French Joint Exercises

Exercise Name (Date)	Context	HQs Involved	Scenario	Other Forces
Perle d'Ouest (Feb 2004)	A bilateral French and Kuwaiti CPX followed by an FTX	N/A	N/A	Unspecified forces
Eolo '04 (Oct 2004)	An EU CPX to validate EU military C2	French FHQ acting as an operational HQ	Peacekeeping operations	None
Golfe '05 (Sep 2005)	A bilateral French and UAE CPX followed by an FTX	French FHQ acting as a multinational operational HQ	Crisis management	The UAE: a combined-arms battalion (-); air and naval forces. France: an armored brigade (-); an amphibious assault group (company sized); air and naval forces
Milex (Nov–Dec 2005)	An EU CPX to validate EU military C2; the sequel to Eolo '04	French FHQ acting as strategic HQ reinforced by a CPCO corps element; German FHQ	Peace-support operations	None
OTIADEX (Dec 2005)	A joint, national French CPX to validate homeland-defense C2 structures	Organisation Territoriale Interarmées de Défense	Homeland defense	None
RECAMP (June 2005)	A joint, multinational EU and AU effort; includes CPXs	N/A	Crisis management	None
Exenau '07 (Feb 2007; annual)	A multinational CPX (afloat)	FHQ, component HQ	Full-spectrum operations	None

SOURCE: Ministère de la Défense, État-Major des Armées, "Exercises Interarmées," Web page, undated.
NOTE: (-) indicates less than one full battalion, brigade, or other group.

hundreds of kilometers into the interior.[43] The purpose of these operations, however, was, presumably, to demonstrate to rebel factions that French forces had the ability to strike them in their base areas, if necessary.

The French Army. As noted earlier, French Army preparations for operations take place in a context in which individuals and units have deep reservoirs of operational experience. Moreover, leaders are relatively senior to and more experienced than their U.S. counterparts. Therefore, these preparations serve more to refresh existing skills acquired during operations than to develop and mature required skills. Many aspects of the French system contribute to the development of adaptability, including the authority and autonomy accorded to leaders at all levels to train their units and the strong emphasis on training commander-leader teams. As in the United States, in France, the French Army's CTC forms the centerpiece of its training effort, and, as in the United States, the CTC is shifting to what the IDA study called "ongoing adaptations through lessons learned,"[44] or, in this monograph's terminology, from a force-preparation model to a force-generation model. Even this shift must be seen in the context of the overall French readiness system, in which collective training is designed to refresh experience and to hedge against an unlikely eruption of combat operations.

Force generation and force preparation take place under the direction of the CFAT. Within the CFAT, the Bureau de Conduite de la Préparation Opérationnelle (BCPO) manages preparation for operations. These activities comprise part of a 16-month readiness cycle that includes operational preparation, an alert (or ready) phase, operational deployment, and a recuperation phase. The BCPO prescribes the major training events for deploying units and key individuals, then coordinates support from the various agencies of the French armed forces and the government. The BCPO also continuously updates this training program based on lessons learned from ongoing operations.

---

[43] See Ministère de la Défense, État-Major des Armées, "Côte d'Ivoire: A l'Ouest, du Nouveau," Web page, July 25, 2007; Ministère de la Défense, État-Major des Armées, "Côte d'Ivoire: Opération 'Wagram,'" Web page, July 23, 2007.

[44] Tillson et al., *Learning to Adapt to Asymmetric Threats*, p. 19.

Compared to the U.S. Army Force Generation model (when it is functioning in its ideal state), the French arrangement occurs in a fairly compressed timeline that is repeated within the lifetime of a unit. The French Army recognizes four types of training: C2, maneuver, live fire (including collective, crew, and individual exercises), and battle hardening (which comprises efforts to acclimatize troops to a specific physical environment, such as mountains or jungles). In practice, the balance of training orients on force preparation (especially for traditional combat operations) at the CPF, although this balance is continuing to evolve. Figure 3.1 depicts the French Army training model.

The emphasis in French Army preparation for operations seems to be shifting from force preparation to force generation. As late as July 2004, French Army leaders apparently saw little need for preparations specific to a given operational environment. As the after-action report

Figure 3.1
The French Army Training Model

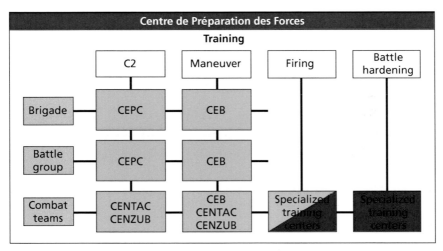

SOURCE: Colas des Francs, "Centre de Préparation des Forces," briefing, August 2, 2007.
RAND *MG836-3.1*

for Opération Licorne, the ongoing French operation in Côte d'Ivoire, put it,

> long periods of pre-deployment training can no longer be justi-
> fied when every unit has been deployed several times to theater,
> when every unit is highly cohesive, and when the rhythm of oper-
> ations imposes a certain economy of men and materials on train-
> ing. A CPX, a brief FTX [field training exercise] on open ground,
> and familiarization training with the equipment to be used on
> campaign seem to represent an optimal solution.[45]

In 2006, however, the French Army stability-operations manual noted a need for preparation specific to the operational environment.[46] Currently, French forces deploying to ongoing operations undergo such specific preparations under the heading of mise en configuration pour la projection [preparation for deployment]. These tailored efforts emphasize academic instruction for commanders and staff but also include a network that links observations from the center for lessons learned (the Division Recherché et Retour d'Expérience) and the training center. (These links are similar to the connections between the U.S. Army's Center for Army Lessons Learned and the CTCs.) Nonetheless, constrained resources and the relatively more benign environment in which French forces operate mean that such efforts take place on a reduced scale.[47]

Still, given the much higher level of resources and intensity devoted to rotations at the CPF, it must be assumed that the French Army as an institution places relatively greater weight on training through such

---

[45] Ministère de la Défense, Armée de Terre, Centre de Doctrine d'Emploi des Forces, Division Recherche et Retour d'Expérience, *Cahier du Retex: Enseignements de l'Opération Licorne*, July 2004, pp. 17–18.

[46] Ministère de la Défense, Armée de Terre, *Doctrine d'Emploi des Forces Terrestres en Stabilization*, pp. 19–20.

[47] Herve Charpentier, email to Stephen Arata, November 7, 2007. General Charpentier currently commands the French Army infantry center and formerly commanded the 9th Brigade Infantry de Marine. Lieutenant Colonel Kelly Marie Carrigg, the U.S. Army Training and Doctrine Command liaison officer to the CDEF, also described this predeployment educational effort in telephone discussions in September–December 2007.

rotations. At U.S. CTCs, commanders tend to follow similar patterns, making practice at the CTC the dominant factor. It is not unreasonable to assume that French commanders tend to make similar choices with regard to training scenarios at their own CTC. The French Army is also considering further consolidation of live-fire training and battle-hardening training under the CPF, as shown in Figure 3.2. Such a consolidation would shift even more of the responsibility and authority for collective training from unit commanders to the CPF.

Once a unit has been identified for commitment to an ongoing operation, the BCPO begins its cognitive preparation for operations under the heading of mise en configuration pour la projection. Experts from the Defense Staff, the Army Staff, the intelligence services, the diplomatic service, and academia orient these leaders to the physical, human, and cognitive terrain of the operation. The École Militaire de Spécialisation de l'Outre-Mer et de l'Étranger (EMSOME) plays a special role, providing up to five days of training in the area of opera-

Figure 3.2
The Proposed French Army Training Model

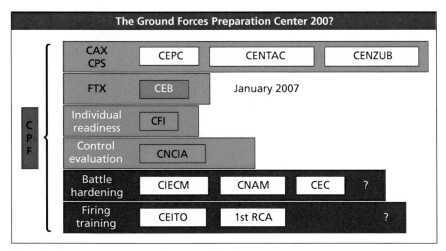

SOURCE: des Francs, "Centre de Préparation des Forces."
RAND MG836-3.2

tions to the officers and the NCOs of the deploying unit. Officers with recent experience in the area of operations provide this training, mostly through lectures.[48] This academic instruction informs the C2 training of the officers of the deploying unit.

C2 training is particularly intensive. In 2005, the French Army conducted 23 exercises for their one corps HQ, 15 exercises for their division HQ, and 78 exercises for brigade HQ. Given that the French have 11 brigades, this works out to a little more than seven CPXs annually per brigade, one of which is an evaluated CPX at the Centre d'Entraînement des Postes de Commandement (CEPC). By way of comparison, U.S. Army Regulation 350-50, *The Army Combat Training Center Program*, prescribes a capacity to conduct 14 division-level and 14 brigade-level evaluated CPXs annually.[49] Many of these exercises are conducted by a single echelon.

Constrained resources, especially the time available for training, have led the French to consolidate the aforementioned C2 and maneuver training under the CPF. As one National Assembly report found in 2002, after a French unit has been deployed on operations for 120 days in a year, it has about 115 days left for training of all sorts (after allowing for leave, weekends, and other inevitable commitments).[50] As the report put it, achieving the statutory objective of 100 days of training annually would require extremely close management of the training calendar. Even assuming that units achieve that objective of 100 days, only about 46 days are typically devoted to training with organic equipment. Moreover, unit commanders tend to be consumed with

---

[48] Discussions with EMSOME officials, Paris, France, March 3, 2008.

[49] Hart, *Avis Présenté au Nom de la Commission de la Défense Nationale et des Forces Armées sur le Projet de Loi de Finances pour 2007 (No. 3341)*; U.S. Department of the Army, Army Regulation 350-50, *The Army Combat Training Center Program*, Washington, D.C.: Headquarters, Department of the Army, January 24, 2003, p. 19. It must be conceded that the number of CPXs presented here includes all CPXs, even the ones whose primary role is to support another HQ by serving either as a subordinate element or as the training unit's higher HQ.

[50] M. Guy Tessier, "Rapport Fait au Nom de la Commission de la Défense Nationale et des Forces Armées sur le Projet de Loi (No. 187) Relatif à la Programmation Militaire pour les Années 2003 à 2008," No. 383, November 25, 2002.

force generation for the next deployment rather than with collective training for high-intensity combat operations. Other resources, such as training space, simulation equipment, and laser engagement systems, are also at a premium. The French Army devotes the equivalent of a brigade's worth of forces to support the CPF.[51]

About two months prior to deployment, units conduct a rotation at the CPF under the task organization in which they will deploy. This training is very much aimed at improving proficiency, and it replaces much of the collective training that units would otherwise have to conduct on their own. This approach is different than that of the U.S. CTCs, which incentivize and enforce unit performance of collective training. In France, training at the CPF is also evolving from its earlier exclusive emphasis on traditional high-intensity combat operations to expanded scenarios that replicate all aspects of the current security environment. As noted earlier, permanent liaisons between active theaters of operation and the training center are supposed to shape training events to reflect the operational environment. By 2007, the French Army recognized that its training resources were ill-suited to modeling complex environments (particularly stability operations taking place under the heading *maîtrise de la violence*). The French are changing both their training centers and their simulations to reflect the contemporary operational environment. They are also adding a brigade training center (the Centre d'Entraînement de Brigade), principally to allow collective live-fire training.[52]

Forces at the company-team level conduct maneuver training in high-intensity combat against a professional opposing force at the Centre d'Entraînement au Combat (CENTAC), and they conduct maneuver training in urban areas at the Centre d'Entraînement aux Actions en Zone Urbaine (CENZUB). Training at CENZUB takes place in both a high-intensity-combat framework and a three-block–war scenario. Although training in the latter case emphasizes tight ROE, CENZUB has only recently begun replicating noncombatants.

---

[51] Tessier, "Rapport Fait au Nom de la Commission de la Défense Nationale et des Forces Armées sur le Projet de Loi (No. 187)," pp. 47–49.

[52] Des Francs, "Centre de Préparation des Forces."

In the second half of 2007, CENTAC opposing forces began to portray both civilians on the battlefield and asymmetric enemies. CENTAC has, therefore, now started to portray training scenarios that include crowd control and training and reaction to peaceful demonstrations.[53] Such training includes a limited degree of multiservice training with the French Air Force.

Finally, the CPF trains company-level leaders and regimental (battalion), brigade, and division (i.e., État-Major de Force) staffs at the CEPC using the Joint Army Navy Uniform Simulation (JANUS) and Simulation de Combat Interarmées pour la Préparation Interactive des Operations (SCIPIO) simulations. Originally designed to model traditional combat (particularly reconnaissance and engagement), neither simulation supported training for stability operations very well. However, the French Army's simulation office has begun adapting these simulations to support training in stability operations. Since summer 2007, the JANUS simulation, which is used to train company commanders and battalion HQ, has modeled urban environments and civil disturbances. In the civil-disturbance simulation, if the player takes the actions that French doctrine prescribes, the scenario improves. If, however, the player responds inappropriately, especially with excessive force, the situation deteriorates, sometimes to the point at which the player unit is overwhelmed.

The CPF uses the SCIPIO simulation to train brigade and division staffs. These exercises can be internally evaluated, conducted at the unit's request, or externally evaluated. In preparation for deployment to an ongoing operation, the CEPC configures the training area to exactly resemble the command post in theater. At present, the SCIPIO simulation models high-intensity combat operations and cannot be used to model peacekeeping or stability operations.[54] However, the simulation office is adapting the SCIPIO simulation. The principal improvements

---

[53] For the recent augmentation of training scenarios, see Kergus, "La Prise en Compte des Spécificités de la Stabilisation dans les Exercices de Préparation et d'Évaluation," p. 47. Moreover, videos on CENZUB's Web site depict traditional urban combat in an environment devoid of noncombatants.

[54] Chary, "SCIPIO V1: Future Training Tool for Major Unit CPs within the French Army," *Objectif Doctrine*, No. 22-02/2001, February 2001, p. 35.

to the simulation include modifying it to allow for deterrence of adversaries and for the simulation of civil-disturbance operations. Currently, the simulation simply models interactions as combat operations that continue until either the friendly or enemy subunit is destroyed. The new version will allow units to react according to their doctrine (either French Army or enemy) and to decline engagement if defeat seems likely. For instance, a terrorist entity detecting a friendly outpost that is properly organized and equipped will decide not to carry out its attack. Crowd behavior will follow the behavior used in the adapted JANUS simulation, in which forces' compliance with the ROE leads to civil compliance. The simulation office foresees fielding this improved version of SCIPIO in 2009. These enhanced simulations may be of interest to the United States.[55]

Battle-hardening training aims more at developing and reinforcing basic military capabilities than at preparing French soldiers for specific operational environments. There are three centers (known as centres d'aguerissement) that conduct battle-hardening training: two for mountain operations and one for commando operations. The descriptions of these centers, especially the mountain-training centers, make it clear that their purpose is principally to train small units and their leaders under challenging conditions. French units train in the mountains because doing so is hard, not because they anticipate fighting there. In that its aim is to place small groups in challenging and unfamiliar settings in order to improve cohesion and adaptability, battle-hardening training is one of the few examples of training that seems to be expressly linked to inculcating adaptability.

French Army leaders have only recently recognized the utility of training tailored to stability operations. According to General Vincent Desportes, the chief of the CDEF, French leaders long believed that training for high-intensity combat also prepared soldiers for stability

---

[55] Discussions with French officers, Fort Belvoir, Va., October 23, 2007; Frédéric Morinière, Guillame Danes, and Laurent Tard, "La Stabilisation: Encore Plus d'Exigence pour les Simulations," *Doctrine*, No. 12, August 2007, pp. 50–52. It is interesting to note that CFAT expects an earlier deployment (sometime in 2009). See Kergus, "La Prise en Compte des Spécificités de la Stabilisation dans les Exercices de Préparation et d'Évaluation," p. 47.

operations and other "lesser included cases."[56] Lessons learned from actual operations, however, indicated that this was not the case and suggested that the two types of operations require significantly different skill sets. For that reason, the French Army has increased its emphasis on training that is specifically oriented on stability operations. In spite of this evolution, however, combat operations still comprise a key element of predeployment training, in no small part because French doctrine presumes that maîtrise de la violence requires the ability to dominate the ladder of escalation.[57]

Although French Army doctrine differentiates between maîtrise de la violence, which the United States equates to stability operations, and coercition, which the United States equates to traditional combat operations, the difference tends to play out more at the operational level than at the tactical level of war. With regard to maîtrise de la violence, French Army doctrine implicitly assumes that French forces will interpose themselves between more-or-less organized parties in conflict. Operationally, the role of French forces is to prevent any belligerent from gaining a monopoly of violence that would preclude a political settlement. In turn, this role primarily requires armed forces to dominate any and, if necessary, all parties to a conflict. In this view, French forces would simply perform the same tactical tasks that they practice at the CPF, but at a different tempo and in a more complex milieu and to achieve a different operational end state.[58]

The interesting questions with regard to French Army preparation for operations concern the rotational cycle and the centraliza-

---

[56] Pengelley, "French Army Transforms to Meet the Challenge of Multirole Future," p. 45.

[57] Pengelley, "French Army Transforms to Meet the Challenge of Multirole Future," p. 45; Tessier, "Rapport Fait au Nom de la Commission de la Défense Nationale et des Forces Armées sur le Projet de Loi (No. 187)," p. 46.

[58] See Ministère de la Défense, Armée de Terre, Centre de Doctrine d'Emploi des Forces, FT-01, secs. 21 and 4.2.2, which list four phases: initial, taking the initiative, mastering violence, and disengagement. The second phase involves separating belligerents and demonstrating French dominance. This is how the situation in the Côte d'Ivoire played out: French forces established a zone of separation between government and rebel forces and held the balance. According to Amnesty International, however, the French forces did not do a very good job of protecting individuals in the zone of separation.

tion and consolidation of collective training. Rotations of HQ staff seem to result in a fairly substantial loss of collective memory, and the four-month cycle means that this loss occurs frequently. In a report on French operations in Côte d'Ivoire, the International Crisis Group noted that the brevity of French deployments inhibits familiarity with the terrain and the population, which contributes to unnecessary collateral damage.[59] It is worth asking, however, whether repetition of the readiness cycle in a given unit improves the unit's effectiveness or degrades it relative to effectiveness of a unit that is both prepared longer and employed longer (as occurs in the U.S. model). The consolidation and centralization of collective training are other interesting features. This consolidation and centralization seem to be the most interesting feature of the French system of preparing for operations, and it would be worthwhile to assess the relative efficiency and efficacy of these features in preparing land forces for real-world operations.

To recap, the French Army employs a CTC model for training, and the model is consolidated to an even greater degree than the U.S. Army's. The French Army seems to be shifting its emphasis from force preparation to force generation. These preparations focus on academic instruction for leaders and staffs. The French Army is beginning to adapt training at the CPF to portray more-complex scenarios, however, and its adaptation of the JANUS and SCIPIO simulations for training leaders and staffs probably merits further assessment and evaluation.

The French Air Force. The French Air Force's collective training mainly supports multinational interoperability, and it places less emphasis on joint operations, either national or multinational. More than the French Army, the French Air Force continues to emphasize traditional combat operations and force preparation. Like the French

---

[59] International Crisis Group, "Côte d'Ivoire: No Peace in Sight," Africa Report No. 82, July 12, 2004. See also Amnesty International, "Côte d'Ivoire: Threats Hang Heavy over the Future," October 13, 2005. It should be noted, however, that another Amnesty International report dealing specifically with this issue demanded an independent inquiry but also seemed to conclude that there was little that French forces could have done to avoid the use of lethal force, given the persistence of government instigated rioters and a paucity of nonlethal equipment. See Amnesty International, "Côte d'Ivoire: Clashes Between Peacekeeping Forces and Civilians: Lessons for the Future," October 5, 2005.

Army, the French Air Force employs a CTC model for training. The interesting aspect of that CTC program is that the CTCs belong to other countries and organizations. The French Air Force participates in the U.S. Red Flag and Air Warrior Exercises, and a total of 50 French air crews were to be trained in 2007. The French Air Force's other major training venues are Canada's Maple Flag (which 67 French crews attend) and the NATO Clean Hunter/Brilliant Arrow exercise (which 80 French crews attend).

The French Air Force participates in a wide range of other multinational exercises. The 2005 NATO Allied Action CPX certified France for a role as the joint-force air component command in NATO's High Readiness Force. Notably, although the French are frequent guests at such major exercises, they seldom host them, thus sparing themselves the need to maintain an expensive CTC infrastructure.[60] Moreover, according to one knowledgeable observer, the French Air Force tends to emphasize air superiority over ground attack, a fact reflected in its participation in foreign exercises, relatively few of which involve ground attack.[61] This is an interesting circumstance given the fact that France does not envision conflict with a near-peer competitor in the foreseeable future.

Finally, as is the case with French Army members, French pilots and flight crews are more senior and have greater experience than their U.S. counterparts. To reduce training costs, the French Air Force concentrates its pilots almost exclusively on flying duty. Moreover, pilots stay in the same squadron and in the same aircraft for seven years or longer. This likely mitigates, to a degree, the requirement for collective training.

The French Navy. The French Navy follows a force-preparation regimen whose contents are governed more by the general characteristics of the maritime environment and the technical capabilities of vessels than by any specific operational environment. According to statute, French vessels must complete certification exercises every two years.

---

[60] See, in particular, Sénat Français, "Projet de Loi de Finances pour 2007: Défense— Préparation et Équipement des Forces: Forces Aériennes," 2006.

[61] Discussions with a French defense analyst, Washington, D.C., September 4, 2007.

The French Navy's four subordinate functional commands for surface warfare, subsurface warfare, aviation, and commandos conduct these exercises. For ships, the exercises follow a pattern of initial qualification in the safe operation of the ship while underway, basic qualification in the operation of mission equipment and systems, and operational qualification in the ships' specific mission areas. Before commencing these exercises, however, each vessel independently conducts intensive individual and collective training in preparation for the exercises. Once the appropriate functional command certifies the unit's qualification, the unit commander maintains a fairly strict training regimen of collective and individual tasks.[62]

Statute further directs that French forces conduct a total of 13 major exercises annually, including four amphibious exercises, three mine-clearing exercises, and two carrier-battle-group exercises. To the maximum extent possible, these exercises are combined with other joint and multinational exercises. The major such exercise is the annual Trident d'Or, a NATO exercise in which the French provide the maritime component command HQ and the bulk of the forces. In 2005, this exercise was a full-spectrum-operations scenario in which Corsica and Sardinia simulated an archipelago in the middle of the Atlantic. Again, the purpose of this annual exercise is more to hone generic skill sets than to prepare French and other naval forces for a specific contingency.[63]

Assessing Training Effectiveness: Opération Licorne, a Case Study

Opération Licorne, the French mission in support of UN forces in the Côte d'Ivoire, is a case study that offers the opportunity to assess French

---

[62] For annual naval training requirements, see Sénat Français, "Projet de Loi, Adopté le 15 Janvier 2003, No. 49." The general model follows the description found in Schank et al., *Finding the Right Balance*, p. 39. For a brief description of an actual series of qualification exercises, see the description of Mise en Condition Opérationnelle '05 in Ministère de la Défense, Marine Nationale, "Mise en Conditon [sic] Opérationnelle dans le Pacifique," Web page, undated.

[63] For unit norms governing French naval training, see Sénat Français, "Projet de Loi, Adopté le 15 Janvier 2003, No. 49." For a description of Trident d'Or '05, see Ministère de la Défense, Marine Nationale, "Mise en Conditon [sic] Opérationnelle dans le Pacifique," Web page, undated.

performance in maîtrise de la violence. The Côte d'Ivoire has been a complex and challenging operational environment for French forces, requiring them to contend with government forces and rebels (all of whom profess intense hostility to France), government-orchestrated riots, and the full gamut of crime and disorder. As with most contemporary French operations, the French Army dominated Licorne, providing the bulk of French forces and the C2 HQ, which was organized around the setup of a standard brigade but modified to function as a tactical-operational HQ. French forces have performed a number of missions, including protecting and evacuating noncombatants; establishing and maintaining separation and equilibrium between the contending parties to allow for the emergence of a political settlement, a task in which they are theoretically acting in support of UN forces; providing humanitarian assistance; and controlling civil disturbances. Overall, French forces have coped with these circumstances quickly and in conformity to their doctrine, indicating a reasonably high level of training effectiveness. French authorities have also expressed a high degree of satisfaction with the high levels of adaptability demonstrated by leaders and units. In 2007, after almost five years of continuous operations, French forces were able to withdraw to their ostensible role as a reaction force from their previously fairly active role in stabilizing the Côte d'Ivoire. The shortcomings that have been revealed appear to be the result of both defects in organization and equipment and the brevity of the forces' rotational cycle rather than the result of shortcomings in French training methods. Still, for all the complexity and challenge of the Côte d'Ivoire, French officials would readily concede that it has been nowhere near as challenging as either Iraq or Afghanistan.

A Brief History of Opération Licorne. The crisis began in September 2002, when rebellion erupted in the Côte d'Ivoire. A movement calling itself the Ivory Coast Patriotic Movement seized control of the north of the country. Ivoirian President Laurent Gbagbo invoked his mutual defense treaty with France, claiming that the rebellion had been instigated by foreign governments, notably Burkina Faso. French authorities placed little credence in this allegation but did feel it necessary to protect and evacuate French citizens.[64]

Fortunately, the French already had a military presence in Côte d'Ivoire, the 43rd Battalion of Marine Infantry, which they augmented with other forward-positioned forces. The operation was initially under the C2 of the 11th Parachute Brigade. During this initial period, French forces also effectively blocked a rebel move to seize the capital. The 11th Parachute Brigade was only the first of many HQ that would control French military operations in Côte d'Ivoire.

By January of 2003, French authorities had facilitated and perhaps compelled the government to reach a cease-fire agreement with the by-now multiplying rebel groups. The Linas Marcoussis Agreement established a zone of separation between the belligerents and was maintained by French and UN peacekeeping forces who were deployed to the Côte d'Ivoire. This period was marked by several sharp clashes between rebel forces and French troops, which generally tended to result in many dead rebels but little overall impact on the conflict.[65]

A new crisis that arose in November 2004 resulted in direct confrontation between French forces under the command of General Henri Poncet on the one hand and the Ivoirian government and President Gbagbo on the other. The January 2003 Linas Marcoussis Agreement had done nothing to resolve the underlying issues, which basically concerned the division of wealth and power between individuals and factions in Côte d'Ivoire, and it had not abated the ambition of any of the belligerents to resolve the issue by force. Rebel groups organized

---

[64] This chronology is derived from Ministère de la Défense, Armée de Terre, Centre de Doctrine d'Emploi des Forces, Division Recherche et Retour d'Expérience, *Cahier du Retex: Enseignements de l'Opération Licorne.* For both the French official skepticism of President Gbagbo's claim of external intervention and their determination to protect their citizens, see M. Eric Raoult, "Rapport Fait au Nom de la Commission des Affaires Étrangères sur la Proposition de Résolution No. 1968, Tendant à la Création d'une Commission d'Enquête sur les Conditions dans Lesquelles le Gouvernement Est Intervenu dans la Crise de Côte d'Ivoire Depuis le 19 Septembre 2002," Assemblée Nationale, Report No. 2032, January 18, 2005.

[65] Ministère de la Défense, Armée de Terre, Centre de Doctrine d'Emploi des Forces, Division Recherche et Retour d'Expérience, *Cahier du Retex: Enseignements de l'Opération Licorne.*

and consolidated under the title of Forces Nouvelles, and the government augmented its capability by buying advanced military equipment and, allegedly, hiring mercenaries to operate it.

By fall 2004, President Gbagbo apparently felt strong enough to abrogate the agreement. While his partisans organized crowds to keep French and UN forces bottled up in their bases, government forces attacked exposed rebel positions and bombed villages that, supposedly, were supporting the rebels. Initially, it seems that General Poncet felt he could do little besides protest these blatant violations of the cease-fire. He was, however, apparently seeking an opportunity for a riposte. When Ivoirian Su-25s bombed the French base at Bouake in November, killing nine soldiers and an American aid worker and wounding 40 more, French forces waited till the planes landed and then destroyed them on the ground. Because of concurrent attacks on French civilians, apparently orchestrated by President Gbagbo's political supporters, French forces also took control of the airport at Abidjan to ensure their ability to evacuate French and foreign nationals.[66]

Both measures, the destruction of Côte d'Ivoire's Air Force and the seizure of the airport, provoked an intense popular reaction stoked by Ivoirian-government propaganda and covertly organized by President Gbagbo's supporters. President Gbagbo's government has maintained to this day that the bombing was an accident, a plausible but unconvincing explanation, given the considerable separation between French installations and rebel positions. In the riots that followed the seizure of the airport, French forces in the city and at the airport were forced to resort to lethal force to protect themselves and maintain control of the airport. At one point, French mechanized infantry surrounded the presidential palace, from which President Gbagbo was presumed to be orchestrating events.

The events of November 2004 marked the apex of the crisis, and tensions have subsided gradually since then. Negotiations between the

---

[66] The description of events in this paragraph and the next summarize the narrative found in International Crisis Group, "Côte d'Ivoire: Le Pire Est Peut-Être à Venir," Africa Report No. 90, March 24, 2005, pp. 8–9. The events are described in even greater detail in Amnesty International, "Côte d'Ivoire: Clashes Between Peacekeeping Forces and Civilians."

Gbagbo government and the Forces Nouvelles led to power-sharing agreements embodied in the Ouagadougou Accords of March 2007. After the agreement was reached, French forces reverted to their role as the reserve for UN forces, and they have largely focused on conducting joint exercises that demonstrate their capability to intervene in any part of the country.

An Assessment of Opération Licorne. French military officials generally express satisfaction with the performance of their forces in Opération Licorne. These forces managed a difficult, complex conflict with minimal French casualties. The July 2004 Opération Licorne lessons-learned report occupies a preeminent place on the CDEF Web site, and a vignette on the events of November 6, 2004, figures prominently in FT-01, *Gagner la Bataille: Conduire à la Paix,* a French Army doctrinal publication of equivalent importance to the U.S. Army's Field Manual (FM) 1, *The Army,* and FM 3-0, *Operations.*[67] Although France's elected officials tend to be less satisfied with the intervention in Côte d'Ivoire, their discontent seems to stem more from the nature of the mission than from the conduct of French forces. French forces have usually mastered relatively quickly the situations in which they find themselves, and they have done so in accordance with their doctrine. This indicates a fairly high level of training effectiveness. Indeed, the lessons-learned report attributes the success of the operation thus far to the high level of leader and collective training. The setbacks that Force Licorne has experienced seem to have resulted more from failures of organization and equipment than from shortcomings in training.[68]

C2 training appears to have compensated for the short rotational cycle to only a limited extent in Opération Licorne, however. On the one hand, French brigade HQ demonstrated an ability to control fairly complex operations without any gaps in effectiveness. Further, brigade staffs adapted themselves relatively easily to the complexities involved

---

[67] Ministère de la Défense, Armée de Terre, Centre de Doctrine d'Emploi des Forces, FT-01, p. 39.

[68] Ministère de la Défense, Armée de Terre, Centre de Doctrine d'Emploi des Forces, Division Recherche et Retour d'Expérience, *Cahier du Retex: Enseignements de l'Opération Licorne,* p. 12.

in serving as a tactical-operational HQ. On the other hand, French training methods do not appear to have transferred situational awareness and understanding successfully between rotating HQ.

To be sure, the bombing of French forces at Bouake may indicate a failure in training. Certainly, French forces had hedged against an air threat after initially deploying forces without air-defense capabilities. By November 2004, however, Force Licorne had at least a rudimentary air-defense capability, and it did not employ that capability when attacked. The fact that they had the capability but did not employ it may indicate that either the staff or the air defenders were not mentally prepared for this eventuality. It may also simply indicate that it was not possible to make the intensely political decision to engage in the compressed time available during the air attack.

French forces' conduct during the November 2004 riots in Abidjan, however, constituted a fairly effective operational and tactical response to an asymmetric attack on the French position in the Côte d'Ivoire. By preemptively seizing the airport, General Poncet averted a situation in which French civilians could, in effect, have been held hostage, an event that would have neutralized French forces. President Gbagbo's subsequent attempts to dislodge the French forces using crowd action faltered because of French firmness. By surrounding the presidential palace with armored forces, the French sent the pointed message that President Gbagbo could not provoke them with impunity and might indeed find himself designated an enemy of France.[69] When President Gbagbo later tried to constrain and intimidate the French with orchestrated riots at the Hotel d'Ivoire and at the airport, French forces demonstrated willingness to use whatever force was required—including lethal force—to maintain their freedom of action.

The French response also demonstrates a high degree of adaptability. When French forces deployed to Côte d'Ivoire in 2002, they did so as at least the nominal ally of the Gbagbo government. One report went so far as to note that the government would have certainly fallen a few months later had French forces not blocked rebels advancing on

---

[69] The French continue to maintain that surrounding the presidential palace resulted from taking a wrong turn, but the move clearly conformed with French doctrine.

the capital.[70] By November 2004, however, the Gbagbo government had become an adversary. It seems likely that the air attack on the French base at Bouake was part of a campaign to neutralize the French while President Gbagbo dealt with the rebels. The French responded rapidly and effectively in this altered situation, however, regaining the initiative in Abidjan and throughout their area of operations. Clearly, General Poncet and his staff did not allow themselves to be surprised in the military sense of being frozen into inactivity.

French troops acted with restraint but used force effectively when threatened. Although human rights groups have criticized the French resort to lethal force, it is important to note that French soldiers opened fire only after considerable provocation, under conditions that threatened their lives and safety, and only on the orders of their leaders. Still, although French forces certainly neither lost control nor got out of hand, there are indications that lower echelons of French command were not quite prepared to cope with the situation. For instance, before French forces opened fire, the Ivoirian crowd managed to drag off at least one French soldier, who was retrieved only with difficulty. Still, the French forces' objective in this situation was to maintain their freedom of action in the face of asymmetric attempts to constrain and intimidate them, a task they accomplished. Apparently, to the French, the death of multiple rioters was an acceptable cost of doing business. Thus, although French forces were unable to prevent a return to violence, they contained it and ensured that they retained control of the ladder of escalation.

The events of November 2004 support this conclusion. As previously noted, a principal aim of maîtrise de la violence is to control the ladder of escalation. Although French forces did little to counter the Ivoirian government's abrogation of the ceasefire, their failure to act was a result of the political reality that they could not, at that time, initiate combat against Ivoirian government forces, whatever their conduct. The air attack on the French forces in Bouake not only gave them a reason to act but also provided a pretext to act in a

---

[70] International Crisis Group, "Côte d'Ivoire: The War Is Not Yet Over," Africa Report No. 72, November 28, 2003.

manner that significantly curtailed the Ivorian government's freedom of action and military capability. When President Gbagbo attempted to respond asymmetrically with orchestrated riots and provocations against French nationals, General Poncet responded, in accordance with French doctrine for controlling mass movements, by "intimidating . . . [local leaders] by the demonstration or employment of force on their combat units."[71] Apparently convinced that neither military "mistakes" nor crowd action would neutralize the French, President Gbagbo acceded to outside mediation.

In the end, however, it is impossible to say more than that Opération Licorne demonstrates that French training methods appear to be adequate for the situations in which French forces are used. This assertion is buttressed by France's strong performance in Afghanistan. The French have successfully navigated complex and difficult situations with a minimum of casualties and without raising public opposition domestically or internationally. That said, for all the complexity of these situations, they do not approach the complexity or the intensity of the operational environment in Iraq. Whether French methods would prove adequate for preparing forces to operate there is difficult to assess.

Training for Coalition Operations

As previously noted, the French presume that most of their major operations will take place in the context of a coalition. The Force HQ at Creil is therefore organized to serve as the possible nucleus of a multinational HQ; officially, the FHQ's language is English, a choice that facilitates multinational interoperability. Additionally, each service maintains at least one HQ capable of serving as a component command in the NATO response force.

Comparison with U.S. Regimes

For the most part, within each domain (i.e., land, air, and sea), French training models and methodologies are fairly similar to those of their U.S. counterparts. However, even more so than in the United States, preparation for operations is a service responsibility in France. Indeed,

---

[71] Ministère de la Défense, PIA 00.200, para. 06-85.

the French lag behind the United States in the development of a meaningful joint-tactical-training capability. Furthermore, the French Air Force is largely integrated into the U.S. training system. As we have observed, much of the French Air Force's collective training takes place in the context of U.S. and NATO exercises, such as Red Flag, Air Warrior, and Maple Flag. In the maritime domain, French training is similar in kind to that used by the United States but is somewhat less intense in degree. Finally, although French Army preparations for operations resemble those of the United States, there are several important differences.

First, the French implicitly rely on operational experience rather than collective training as the foundation of unit capability. Both units and soldiers have been spending a quarter of their time in deployed operations for more than a decade. Consequently, unit collective training and even CTC rotations tend to be less intense. As noted earlier, the culminating predeployment training exercise for a French regiment (i.e., a battalion combat team) typically lasts about two weeks and serves more to refresh existing capabilities than to augment them. French officials do see collective training as important, but they rely relatively more heavily on educating leaders and providing C2 training than on conducting maneuver training. At the brigade level, French commanders and staffs conduct roughly three to four times as many evaluated CPXs as their U.S. counterparts.

Second, French Army operational rotations are both shorter and much more frequent than those of the U.S. Army and the USMC. The French practice of rotating units every four months has a significant impact on continuity of operations in theater. It does seem, however, to enhance units' and soldiers' ability to adapt. On the whole, these two factors seem to cancel each other out.

Third, although the French, like the Americans, employ resource-intensive CTCs, the French CTCs have increasingly functioned more as a substitute for unit collective training than as such training's culminating event. In fact, the French Army has explicitly made the decision to pursue this model, and, to the extent that one considers French Army units effective, this decision seems to be justified.

The similarities between the French and U.S. systems outweigh the differences, however. In spite of an explicit commitment to the principle of adaptability, the French have found it necessary to tailor training to the operational environment. They have developed a system for integrating lessons learned from ongoing operations into training on a near-real-time basis, and they are modifying scenarios and capabilities for both maneuver training and C2 training to support these training scenarios in a manner very similar to that used by the U.S. Army. Opposing forces at the CPF now replicate terrorists, civil disturbances, and a range of other irregular-warfare features. Computer simulations will increasingly model the behavior of civilian populations, terrorists, and insurgents.

The French Army's training establishment, however, continues to prepare French units for a relatively narrow range of contingencies. The French Army, like the U.S. Army, is shifting its training focus toward the middle of the spectrum of operations, even as it struggles to maintain its proficiency in conventional combat operations. In practice, this shift means that French training is now largely oriented on stability operations in an African context. France's training infrastructure does not support generating forces for a set of significantly varied contingencies.

Adaptability Training

Reflecting the general ethos of the French military training system, the French armed forces emphasize operational experience and leader development as their primary vehicles for inculcating adaptability. Collective training plays a role, but indirectly. C2 training helps foster the development of effective commander-leader teams, an element the IDA study identified as important in fostering adaptability. The substantial autonomy afforded to individual commanders in managing their training seems, incidentally, to develop their adaptability as well. Neither C2 training nor autonomy seems explicitly or deliberately designed to foster individual or collective adaptability, however.

The French armed forces attach extraordinary importance to the individual and collective qualities of réactivité and polyvalence, roughly translated as adaptability and versatility, respectively. Indeed, inculcating adaptability is one of the principal goals of French professional military education. The French attach considerable importance to the ability to adapt to foreign cultural contexts. Although few, if any, aspects of the French training-and-readiness model are explicitly connected with the development of adaptability, several fall within the general categories articulated in the IDA study. Most importantly, frequent, brief overseas deployments force French military individuals and organizations to adapt to unfamiliar situations and conditions on a regular basis, albeit in a narrower range of locations than the range of contexts faced by U.S. forces. Second, French professional military education emphasizes the inculcation of initiative and adaptability at all levels. Third, the French armed forces, especially the French Army, place a heavy emphasis on training commander-leader teams, conducting an average of seven brigade CPXs annually.

On the other hand, French collective training, including CPXs, focuses mostly on preparing units to perform anticipated missions under known conditions. The IDA study characterized this kind of effort as "ongoing adaptation through lessons learned."[72] Much like the U.S. armed forces, the French Army CTC has shifted its emphasis from high-intensity combat to full-spectrum operations in a complex operational environment. Still, even as France's training environment has become increasingly focused on ongoing operations, the French have continued to emphasize breadth of capability. For instance, their training center for urban operations, CENZUB, presents three major urban patterns (industrial, market, and center-city, including multistory buildings). French authorities believe that requiring commanders and units to confront these different environments in a compressed timeframe promotes the ability to adapt to a variety of urban environments.[73]

---

[72] Tillson et al., *Learning to Adapt to Asymmetric Threats*, p. 19.

[73] Nicolas Tachon, "Educating for Military Operations in Urban Terrain," briefing to RAND Corporation staff, Tours, France, March 4, 2008. See also Kergus, "La Prise en

Finally, it must be remembered that collective training is only one aspect of the French model for generating military capability (including adaptability) and that it is not necessarily the most important one. In discussions, French officers repeatedly emphasized the role of operational experience in expanding the adaptability and cultural understanding of leaders and soldiers.[74] To the extent that the French military succeeds in inculcating adaptability, it does so more through the way it employs service members than through the way it trains them. Professional education also plays an important role, both in exposing leaders to other cultures and in thoroughly grounding leaders in the technical and tactical fundamentals of their profession.

Operational Experience: The Best Teacher

While deployed, French junior leaders experience considerable autonomy. Junior officers and NCOs frequently find themselves in important posts and separated from company and regimental HQ by some distance. Relatively junior air officers may find themselves serving as the air component commander of a small joint task force. In these distributed operations, geographically isolated French forces must not only make their own decisions but also find their own sources of supply. "Adapt or go hungry" seems to function quite effectively as an imperative for French soldiers, sailors, and airmen.[75]

Collective Training: An Indirect Contribution

Collective training plays mostly an indirect role in the French Army's efforts to inculcate adaptability. For instance, the frequency and scope of its CPXs conform closely, but not completely, to the IDA study's prescriptions for training commander-leader teams. Still, it is not clear how effectively this training prepares HQ staffs for the missions they will undertake. General de Brigade Frank LeBot, who commanded

---

Compte des Spécificités de la Stabilisation dans les Exercices de Préparation et d'Évaluation," p. 48.

[74] Discussions with French officers, Washington, D.C., September 25, 2007, and Tours, France, March 4, 2008.

[75] Discussions with French officers, Washington, D.C., September 25, 2007.

operations during Opération Licorne, noted that every rotation of HQ staff seemed to involve a complete loss of collective memory.[76]

In fact, training management probably plays a larger role in inculcating adaptability in French commanders than does collective training itself. Although the BCPO determines most of the training program for deploying units, unit commanders are given substantial autonomy in the design and conduct of training events. And although the BCPO requires that higher HQ certify subordinate units' readiness, the officers with whom we spoke felt that unit commanders actually made that critical assessment themselves. According to Leonard Wong, a leader's freedom to develop his or her unit's training program is an important element in developing independent judgment and is related to adaptability.[77] The French officers with whom we spoke considered units' independent collective training at least as important, if not more so, as training conducted at their CTC. Indeed, the capstone predeployment exercise is usually a unit field-training exercise whose conditions and conduct are entirely the responsibility of the regimental commander.[78]

Professional Military Education
Almost all of the French officers with whom we spoke emphasized how important adaptability is to the French military, and almost all cited education at their military academies as a key element in developing that attribute.[79] Officer development at the military academies empha-

---

[76] Frank LeBot, "Licorne, or the Challenge to Reality," *Doctrine*, No. 9, June 2006.

[77] Leonard Wong, *Stifled Innovation? Developing Tomorrow's Leaders Today*, Carlisle, Pa.: Strategic Studies Institute, 2002.

[78] Discussions with French officers, Washington, D.C., September 25, 2007.

[79] The French military academies include l'École d'Air, l'École Navale, l'École Spéciale Militaire (St. Cyr), and l'École Militaire Interarmées. The École Militaire Interarmées is the French equivalent of Officer Candidate School. It is a two-year program offered to enlisted soldiers who possess the French baccalaureate and two additional years of education, the level of qualification necessary to enter St. Cyr directly from civilian life. The École Militaire Interarmées owes the shorter length of its curriculum to its students' high level of military experience; military training consumes one full year of the three years a student spends at St. Cyr.

sizes adaptability—including cultural awareness—as one of two key attributes of the future officer. In fact, this educational emphasis on adaptability extends across all ranks and specifically includes NCOs. Begun in 2002, the "ESM [École Spéciale Militaire] 2002" reform of the curriculum at St. Cyr has emphasized more seminar instruction, foreign-language education, and independent study, including completion of an independent research project lasting three months in a foreign business, laboratory, or educational institution. The point of these educational reforms is not to prepare officers for a specific culture but rather to sensitize them to the existence of cultural differences and to prepare them to adapt. For instance, the independent study not only sharpens academic critical and creative thinking skills but also contributes to cadets' adaptability by forcing them to make their own living arrangements in a foreign context. The French Navy and Air Force academies have largely followed suit, although they do not send as many officers overseas for independent study.[80]

That said, one should be cautious in accepting French military education as a model for inculcating adaptability. One French officer who taught at the United States Military Academy and is currently on staff at St. Cyr asserted that there are no significant differences between the curriculum and pedagogy used at the United States Military Academy and the curriculum and pedagogy used at St. Cyr. Indeed, many of the ESM 2002 reforms, including seminar classes and independent projects, have long been staples at the United States Military Academy. Further, although the emphasis on language instruction at St. Cyr is quite strong (cadets are required to learn English and another language), most languages studied are European: Few cadets study Arabic or Chinese. What may differ between the two schools, however, is that

---

[80] Telephone discussions with French Army officers, September 25, 2007, October 24, 2007, and October 29, 2007. See also Bernard Boëne, Thierry Nogues, and Saïd Haddad, "À Missions Nouvelles des Armées, Formation Nouvelle des Officiers des Armes? Enquête sur l'Adaptation de la Formation Initiale des Officiers des Armes aux Missions d'Après-Guerre Froide et à la Professionnalisation," Centre d'Études en Sciences Sociales de la Défense, 2001, pp. 25–27, 79–85.

the French military's educational system places explicit importance on adaptability as a desired attribute.[81]

The Train, Advise, and Assist Mission in France

In addition to investigating predeployment training for operational missions, we were also asked to consider the manner in which French forces are trained to conduct advisory and training missions in other countries. France is an interesting case in that it has a relatively large footprint abroad when it comes to building the capacity of less-capable partner countries around the world. This section considers the following aspects of France's TAA missions:

- the selection of advisers and trainers
- how organizations conduct the training
- which specific skills are trained
- where the trainers are deployed
- how training is assessed and the nature of the lessons-learned process
- key distinguishing features.

The Selection of Advisers and Trainers

The primary objectives of French military assistance and advice are often diplomatic rather than military and are intended to strengthen ties between France and the country whose forces are being advised. For these reasons, special care is taken to ensure that French advisers are knowledgeable about their operational environment. EMSOME prepares these officers for duty as advisers and as attachés with an intensive academic course of instruction tailored to the country to which the

---

[81] Telephone discussions with French Army officers, October 29, 2007; Boëne, Nogues, and Haddad, "À Missions Nouvelles des Armées, Formation Nouvelle des Officiers des Armes?" p. 85.

prospective advisers are going. Instructors at EMSOME are recent veterans of similar duty.[82]

The strictly military aspect of training is, however, less important. A retired French officer with whom we spoke noted that although foreign armies appeared to value the training they received, they tended to disregard it in practice in favor of deeply ingrained, indigenous patterns of operation. Thus, the key goal is customer satisfaction rather than the improvement of indigenous capability. That said, the French Navy has an interesting program for preparing sailors to train foreign navies. The French Navy maintains a database that tracks the capabilities and equipment of foreign navies. Using that database, French personnel can identify partners' possible training requirements. French trainers then learn how to train partners in the use of relevant equipment, which may be neither French nor particularly modern, through tailored distance-learning programs.[83]

It is important to note the high priority that the French Army places on advisory duty. To begin with, advisers are carefully selected, and advisory duty is a key discriminator in selection to flag rank. Moreover, French advisers typically have at least as much experience as the partner forces they are training (e.g., former battalion commanders advise battalion commanders). Finally, tours of duty are three years long, which facilitates rapport and ensures that advisers become thoroughly grounded in the local environment. It was beyond the scope of this study to assess the opportunity costs associated with placing such priority on advisory duty. What does seem clear is that the French believe advisory duty is very important and, apparently, are willing to fill these billets at the expense of other assignments.

In France, there are two tracks—advisers and training teams—for conducting TAA missions abroad. The selection process for both tracks is fairly rigorous. Advisers are often embedded in the host nation's ministry of defense and typically serve in that role for one year. They are selected after an in-depth interview with a jury, which consists of a

---

[82] École Militaire de Spécialisation de l'Outre-Mer et de l'Étranger, home page.

[83] "NAVFCO—The French Navy's Arm for the Training of Friendly Foreign Navies," *Asian Defence Journal*, December 1999, pp. 44–45.

committee chair, a psychologist, and a former adviser. The ideal advisers are deemed to be those who know the destination country well and those who have (preferably recent) experience working as an adviser.[84] Therefore, advisers are often selected for more than one mission over the length of their careers. Once deployed, advisers are under the supervision of the French defense attaché and report to their home battalion commander for combat support, but their salaries are paid by the French Ministry of Foreign Affairs. The idea is to try to disconnect the advisers as much as possible from the French Ministry of Defense, at least in the host nation's eyes, so that the advisers are seen as providing objective advice to the host nation. A key point is that TAA missions are seen as career enhancing, and supplying advisers is considered a part of a French battalion's mission. Each year, 20–30 advisers are selected from each French combat-arms battalion. After a year-long deployment, the advisers are typically given command of a battalion.

Training-team positions differ from advisory positions in a number of ways. First, training-team missions are typically shorter than advisory positions. Second, trainers are under the operational command of the French Ministry of Defense while in theater. Third, trainer salaries are paid by the French Ministry of Defense. However, trainers are selected through a process similar to that used to select advisors (i.e., through a jury, a psychologist, interviews, etc.).

How Organizations Conduct the Training and Skills Taught

EMSOME, which was founded in 1901 to support the Foreign Legion, is the main venue for training French and foreign military training teams and advisers. Ninety percent of all deploying personnel are trained by EMSOME. Each of the 20 EMSOME instructors is an active-duty military member, although academics are brought in as guest lecturers for courses on the Middle East and Islam.[85] Figure 3.3 shows the overall breakdown of EMSOME training responsibilities.

---

[84] It is worth noting that the French personnel system tracks individuals with prior advisory experience and language capabilities.

[85] Discussions with EMSOME officials, Paris, March 4, 2008.

Figure 3.3
EMSOME Training Responsibilities

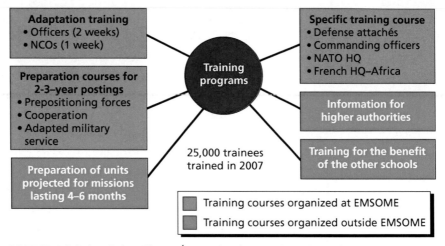

SOURCE: Ministère de la Défense, État-Major des Armées, "Carte des OPEX [Opérations Extérieure]," Web page, undated.
RAND *MG836-3.3*

The training of those deploying to missions abroad below the battalion commander level takes place at the EMSOME facility in Paris. The course is two weeks in duration. Training for battalion commanders typically takes place postdeployment and in country due to time constraints.

The focus of TAA-mission training is primarily on training the right behavior. The following concepts are emphasized in the TAA curriculum under the rubric of "knowing (the situation), knowing how to be, and knowing how to act:"

- Do not patronize.
- Do not underestimate.
- Do not foster conflict.
- Do be empathetic.[86]

---

86 Discussions with EMSOME officials, Paris, March 4, 2008.

French advisers, especially those serving as part of a training team, are taught to never increase the level of violence but rather to try to defuse any bad situation. However, redlines are established with foreign countries, and French forces communicate that they will escalate if these lines are crossed.[87]

Adaptability in TAA missions is a topic emphasized for French trainers at EMSOME. A cornerstone of EMSOME's training involves inviting those with recent operational experience in the host nation to speak to the deploying trainers and advisers. During deployments, emphasis is placed on minimizing reliance on support structures in France. For example, platoon leaders are often placed in situations where they are 500 km from support. For the French, training for adaptability is about practical experience. During TAA missions, decisions are made at the lowest level possible. Additionally, trainers are generally not penalized for minor errors; rather, they are given the freedom to adapt to the situation and the freedom to learn from their mistakes.[88] Adaptive training focuses on

- military humanism
- open-mindedness
- gaining a better understanding of human feeling
- gaining confidence in decisionmaking and risk taking
- learning to show humility by listening rather than speaking during the early days of a deployment to a host nation
- integrating with the local population as much as possible.

Although EMSOME's students are predominately French, about 12 percent of the students are officials from foreign countries (e.g., the UK, Germany, Austria, Poland, Slovakia, and other EU countries) who are bound for Africa. Countries that can afford to send their students to EMSOME do so, and students from countries (e.g., Cameroon and Senegal) deemed important to French national strategic interests who lack the necessary resources are sponsored at ENSOME by the French

---

[87] Discussions with EMSOME officials, Paris, March 4, 2008.

[88] Discussions with former French adviser, Washington, D.C., January 2008.

government. Those who demonstrate the potential to support French models are selected to attend EMSOME by in-country French defense attachés.

Where Trainers and Advisers Are Deployed

France deploys its advisers and training teams only to places where it has a national interest. France categorizes TAA-mission partners into three levels. At the top level, which is reserved for militarily advanced allies, France's goal is to improve interoperability, and the focus is on exercises and simulation training. Examples of partners at this level are the United States, the UK, and Germany. At the second level, the goal is to develop the European defense industry, and the main partners are, therefore, other EU members. At the third level, the level at which most TAA missions take place, the goal is to promulgate France's worldview abroad.[89]

French TAA missions are primarily focused on Francophone Africa, the Balkans, Afghanistan, and France's former colonial territories in the Caribbean. Figure 3.4 shows the destinations of both operational missions and TAA missions. Commitments at the time of writing, including the prepositioned forces and units committed in operations, total about 35,000 troops, of which 23,000 come from the French Army.

Approximately 7,000 French troops are permanently stationed in Africa. The Reinforcement of African Capabilities to Maintain Peace (RECAMP) program trains individuals and units for African peace-keeping operations. RECAMP operations to date include the Economic Community of West African States (ECOWAS) 1998, the Economic Community of Central African States 2000, the South African Development Community 2002, and ECOWAS 2005. RECAMP training consists of three operational components: political-military seminars, staff exercises, and field exercises. RECAMP claims to cooperate closely with the UN, the Organization of African Unity, and such subregional organizations as the South African Development Community, ECOWAS, the Council for Peace and Security in Central Africa,

---

[89] Discussions with officials at the Embassy of France, Washington, D.C., April 2007.

Figure 3.4
French Military Destinations

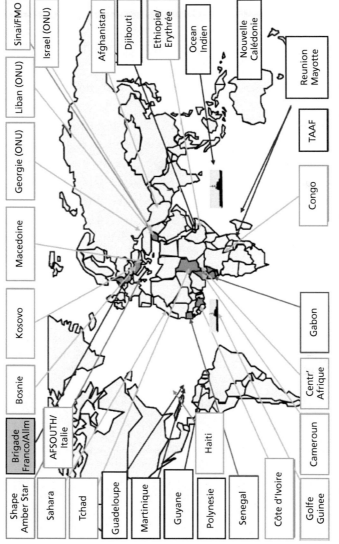

Shape
Amber Star

Sahara

Tchad

Brigade
Franco/Allm

AFSOUTH/
Italie

Guadeloupe

Martinique

Guyane

Polynesie

Senegal

Côte d'Ivoire

Bosnie

Kosovo

Macedoine

Georgie (ONU)

Liban (ONU)

Sinai/FMO

Israel (ONU)

Afghanistan

Djibouti

Ethiopie/
Erythrée

Ocean
Indien

Nouvelle
Calédonie

Reunion
Mayotte

TAAF

Congo

Gabon

Centr'
Afrique

Cameroun

Golfe
Guinee

Haiti

SOURCE: École Militaire de Spécialisation de l'Outre-Mer et de l'Étranger, "Command Briefing," March 3, 2008.

RAND *MG836-3.4*

and the Intergovernmental Authority on Development. RECAMP's focus is increasingly multilateral, and its main partners are the EU, the United States, and Canada. Moreover, funding for RECAMP is being increasingly incorporated into the EU's aid system and into international financial institutions. The totals for TAA in particular, however, are considerably smaller. EMSOME reported as complete the training missions shown in Figure 3.5.

How Training Is Assessed and the Nature of the Lesson-Learned Process

In France, reports from the field emphasize the problems and challenges, but the lessons-learned process appears to be only loosely connected to EMSOME. Moreover, very little (if any) analysis or validation of lessons takes place because of manpower limitations, and there are no metrics for assessing the effectiveness of the training provided to the advisers and training teams. It appears that the assessment of TAA missions is largely based on anecdotal evidence, such as interviews with

Figure 3.5
EMSOME's Completed Training Missions

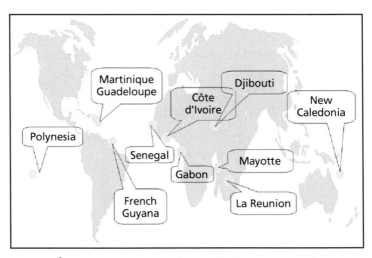

SOURCE: École Militaire de Spécialisation de l'Outre-Mer et de l'Étranger, "Command Briefing."
RAND MG836-3.5

officers with recent operational experience. The key issue for the French is whether France is working with the right countries to support its national interests.[90]

Key Distinguishing Features of the French TAA Approach

As previously mentioned, there are some distinguishing features of the French approach to TAA missions. First, French advisers are embedded in the host-nation ministry-of-defense structure; French trainers try to blend in by wearing the local uniform for the duration of their training mission. Second, predeployment training is intended to encourage further study, not to be an end in and of itself (as evidenced by the fact that some predeployment courses last only six hours). Third, the French focus on training that provides a perspective of the destination host nation through a military lens. Fourth, in some destination host nations, such as Djibouti, there is a great deal of logistic support, but in others, such as nations in Central Africa, the trainer is very much alone. Finally, France only deploys its advisers and trainers to regions of national interest.

Key Insights

In summary, the French armed forces employ training models and methodologies similar to those used by the U.S. military, and they do so for similar reasons. The French share the U.S. assessment of the current and future security environment, and their doctrine describes a very similar range of military operations. Like the United States, France uses a CTC model; in France, the CTC is consolidated at the CPF. Unlike the U.S. model, however, French training at the CPF aims more explicitly at improving performance and is not targeted at validating preparatory training. The French, like the Americans, are moving from a force-preparation model to a force-generation model, a decision based on their experience in sustaining complex contingency opera-

---

[90] The French Lessons Learned Center is reportedly looking to the U.S. Army Center for Lessons Learned (CALL) for ways to streamline the process.

tions, primarily in the Côte d'Ivoire. In France, however, this transition is occurring at a slower pace than in the United States because it is based on a relatively less demanding operational context. Responsibility for force generation falls even more heavily on the services in France than it does in the United States because there are no permanently constituted joint HQ below the level of the FHQ at Creil, and, therefore, there are no permanently organized joint forces. This fact, combined with the expectation that HQ and forces will be tailored to each specific contingency, seem to imply an increased requirement for predeployment training.

In the French view, deep reservoirs of experience (afforded by frequent rotations to the same locations), the small size of the French military, and the country's greater emphasis on officer education mitigate the need for predeployment training. The French military also places greater emphasis on command-post training than on collective training because of constrained resources, especially training land. The French training model seems adequate to meet the need to prepare soldiers for complex operations in an irregular-conflict environment.

Consequently, the primary conclusion to be drawn from this analysis is that the French have found neither a dramatically different nor a dramatically more effective model for training their forces for operations. The core of their training model is still a CTC in which trainers are seeking to replicate the operational environment with increasing verisimilitude. The importance of France's CTC (the CPF) appears to be increasing with each passing year as the French increasingly consolidate collective training under its aegis. There are five insights to be derived from the study of French training models and methodologies. First, the totality of the French system, including operational experience, leader development, and training, is at least somewhat effective in inculcating adaptability. France's emphasis on adaptability and implementation of methods for inculcating it seem to be effective. It is important to remember, however, that the French system is one in which aspects of operational experience and professional development appear to outweigh collective training. It is difficult, therefore, to assess whether these educational and training methods actually result in increased adaptability or whether increased adaptability results in increased effectiveness. Certainly, French forces have been operationally effective, albeit in the relatively familiar cultural context of francophone Africa.

Second, it bears special mention that the most important aspect of this system for inculcating adaptability is operational experience. Put simply, the entire French system of force development and operations assumes the existence of this quality and depends on it. Junior officers and NCOs are placed in charge of isolated posts or entire regions without detailed guidance. Units, both ashore and afloat, are often responsible for improvising their own logistic support. Moreover, units and leaders have a great deal of practical experience. Furthermore, both French officers and NCOs are, on average, three to four years older than their U.S. counterparts and are generally commanding at least one rank higher relative to their position. Pilots are older, have accumulated considerably more flying time, and so on. Finally, commanders have almost complete autonomy in training their units, a circumstance correlated earlier in this monograph with the development of adaptability.

Third, in spite of their long-standing reliance on the adaptability and versatility of leaders and soldiers, the French armed forces are finding that adaptability does not sufficiently prepare units for the complex operational environments they are facing today. Both French Army doctrine for stability operations and training at the CPF are increasingly recognizing a need to tailor training to a specific operational environment. In effect, the entire French training establishment is reorienting to the conditions found in francophone Africa.

Fourth, using CTCs early in the training cycle appears to work. According to the U.S. Army model, training at a CTC should build on and validate extensive collective training conducted by the unit at its home station. In the French model, training at the CPF explicitly replaces much of that home-station training. The French have chosen this model in no small part because of their much shorter readiness cycle, which allocates only four months to prepare for deployments. Moreover, they have significantly fewer units for their one training center than does the U.S. Army for its three maneuver training centers; along with France's reduced geographic scale, the reduced number of units makes it more feasible for the French to operate in this manner. Nevertheless, the French emphasis on using CTCs as training vehicles rather than as "finishing schools" is a model worth considering.

Fifth, the French system leverages both experience gained during operations and leader education and training. The French Army places more emphasis on C2 training than does the U.S. Army, and it places less emphasis on maneuver training. It conducts almost twice as many evaluated CPXs as does the U.S. Army Battle Command Training Program, but it conducts no maneuver training above the company level. The cost and availability of training resources dictate this proportion, but French HQ have demonstrated the ability to handle complex operations at very low levels. This approach may also be worth U.S. consideration. Within C2 training, the French appear to use academic instruction as their primary force-generation vehicle rather than attempting to actually replicate the intended operational environment during CPXs.

Finally, it is clear from our analysis that the French armed forces have proven both effective and adaptable in complex and challenging operational environments. Those environments, however, have mostly been in francophone Africa or the Levant—places where the French have been operating for over a century. Nevertheless, their recent assignment to Afghanistan offers the opportunity to see how their methods prepare forces for this more lethal environment, and this experience should be examined. Finally, it was not possible to assess French forces' performance in a highly lethal environment, such as Iraq, because no such case exists. Therefore, we were unable to evaluate French training methods for that kind of contingency.

CHAPTER FOUR

The United Kingdom

Introduction

The UK has struggled over the past 25 years to balance the demands it places on its armed forces with the resources it has made available for readiness and operations. In part, this struggle has existed because the UK has chosen to participate with other countries and organizations in attempts to influence events around the world. There have been some notable occasions, however, when the actions of others have, in the minds of British decisionmakers, demanded a military response. During this time, the way in which the British armed forces prepare for and execute operations has evolved considerably; now, these forces are almost wholly focused on supporting deployed, joint operations.

Since the early 1980s, the British armed forces have moved from an administrative and operational structure dominated by the single services acting in roles little changed since the end of the Second World War to the present condition of single services focused on delivering operational capability through highly joint structures. This change has included the adoption of significant joint administrative functions to replace those that previously resided in the individual services. Nevertheless, the services have retained their strong identities. The forces are highly deployable and flexible and are able to participate across the full range of combat and noncombat missions demanded of modern military forces. The capabilities of the British armed forces are the result of continuous high-level planning whose purpose is to address, with constrained national resources, the challenges presented by world events. The 1998 *Strategic Defence Review* (SDR)—the most recent review of

British defense, discussed in more detail later in this chapter—built on the foundations of almost two decades of defense-policy reviews and military operations. Without these foundations, the SDR might not have delivered (or attempted to deliver) the forces and structures that now permit the UK to participate as readily and effectively as it does.

Why does the UK focus on deployed operations, and how do the single services work individually to deliver joint forces for joint C2? A comparison of the British armed forces with the militaries of the United States or other nations will mean more if this symbiotic, joint–single-service relationship is understood. In this chapter, we examine in detail the policies that have guided the UK for the last three decades; the ways in which the armed forces have changed, both in light of world events and in response to allocated resources; and how the forces now prepare for and undertake deployed operations. A summary of this process of accelerating change coupled with concurrent high levels of operational commitment—a process that has dominated the shaping of the British armed forces—will provide a deeper background into those forces' current situation and practices. From this background, it may be possible to gain a greater understanding of what may or may not work for U.S. forces. In a number of cases, the UK may have adopted U.S. practices and adapted them to its own needs; such practices may be of particular interest.

The closeness of the trans-Atlantic relationship between the UK and the United States means that many of the ideas and practices of each nation are known by and shared with the other, but the Royal Navy (RN),[1] the British Army, and the Royal Air Force (RAF) are not smaller copies of their U.S. counterparts. The UK has developed and put into practice a joint approach to operations that many other nations, such as France, are now emulating. In some respects, the smaller size of the individual British services has aided the evolution of a more "joined-up" way of training, planning, and operating. There are other features, however, that make the process work, such as a uniform system of describing the readiness of units. As a result, all British operations are now joint: A force of the required size for a given operation can be assembled quickly, and the single-service components are

---

[1]   The Royal Marines are part of the RN.

able to operate together effectively. The British armed forces are experiencing problems, however, especially in light of the tempo of deployed operations as of the time of writing. Throughout 2007 and the early part of 2008, there was particularly active debate in the UK about whether resources were out of balance with the operational demand and so whether there was undue stress on the armed forces.[2] These problems are not caused by joint structures and processes but rather by the imbalance between the resources committed to the forces and the demands placed on them.

As described in the next section, the UK has placed NATO at the heart of its defense policy. In addition to the strategic benefits of doing so, there have been operational benefits: British doctrine has been developed alongside that of NATO, NATO exercises have enabled British units to practice procedures and command at almost all levels, and those same exercises have provided vital training resources (for example, opposition forces) that have enhanced unit preparation for operations and been used as a method of measuring readiness.

The sources we consulted for this chapter include a wide variety of online resources, government reports, and interviews with members of the British defense establishment.

Defense Reviews and Reality

The UK has conducted eight defense reviews (including one every ten years until 1990) since 1945. In the 1990s, there was a process of almost continuous review until the publication of the 1998 SDR. A minireview called *The New Chapter* was undertaken in 2002 after the 2001 terrorist attacks on the United States. These nine reviews are summarized in Table 4.1.[3]

---

[2]   See, for example, the comments of David Crausby, MP, and others in the House of Commons' Foreign Affairs and Defence Debate of November 12, 2007 (House of Commons, "Statements in the House of Commons," November 12, 2007, *Parliamentary Debates*, Commons, 5th ser., Vol. 467, col. 446) or those of James Arbuthnot, MP, during the debate following introduction of the Armed Forces Personnel Bill of January 10, 2008 (House of Commons, "Statements in the House of Commons," January 10, 2008, *Parliamentary Debates*, Commons, 5th ser., Vol. 470, col. 601).

[3]   It was traditional in the UK to name reviews after the incumbent Secretary of State for Defence even though the announcements were sometimes made in the course of normal

Table 4.1
British Defense Reviews

Year	Title
1957	*The Sandys Review*
1967–1968	*The Healey Review*
1975	*The Mason Review*
1981	*The Nott Review*
1990–1992	*Options for Change*
1993	*The Rifkind Mini-Review*
1994–1996	*Defence Cost Studies: Front Line First*
1998	*The Strategic Defence Review*
2002	*The New Chapter*

Each of these reviews has, in one way or another, affected the UK's armed forces. Some reviews, such as the 1998 SDR, have described how the government of the day sees the UK's role in the world and what contribution the armed forces should make; others, such as *Front Line First*, have tried to reduce costs while maintaining capability. Since *The Nott Review* of 1981, the reviews have had a cumulative effect that has molded the UK's single services to deliver a more effective joint capability. On occasions, such as the impact of the Falklands Campaign on *The Nott Review*, the outcome has been in spite of the then-government's original intentions.

It is the reviews and military operations since 1981 that have perhaps had the greatest effect on the current capabilities of the UK's armed forces. The following paragraphs describe each of the reviews (and concurrent major military operations) conducted between the early 1980s and 1998.[4]

---

business with the publication of annual departmental statements, such as those reporting defense estimates.

[4] The UK armed forces were significantly committed to an operation, called Operation Banner, in Northern Ireland throughout this period. For the most part, this role fell to the British Army; approximately 13,500 soldiers, mainly infantry, supported the then–Royal

The Nott Review (1981).[5] Conducted by a new government against the backdrop of a Soviet military buildup and economic constraint in the UK, this review was a re-entrenchment designed to consolidate both the UK's defense capabilities and the country's NATO commitments. It confirmed that the Trident system (provided by the United States) was to be the UK's strategic nuclear deterrent, and it included provisions for the rejuvenation of the British Army of the Rhine. To pay for these measures, there were to be cuts to the RN (notably, disposal of one of the service's three small aircraft carriers and both of its amphibious ships).

The Falklands Conflict (1982). The lessons of this conflict were assessed quickly and published in December 1982;[6] the cuts to the RN planned by *The Nott Review* were cancelled. It was acknowledged that the UK needed armed forces with "flexibility, mobility and readiness . . . for operations in support of NATO and elsewhere."[7]

Options for Change (1990–1992).[8] The government initiated a series of studies in 1990 to restructure the UK's forces in light of the fall of the Berlin Wall in 1989. The studies, collectively known as *Options for Change*, were to deliver the peace dividend by reducing defense expenditures as a percentage of gross domestic product (GDP). Despite cuts to static forces (for example, the decision to halve the British Army of the Rhine), the government's intention was to maintain "smaller forces . . . [that were] better equipped, properly trained . . . flexible and mobile and able to contribute both in NATO and, if necessary, elsewhere."[9]

Ulster Constabulary, although the nature of this role has changed since the Good Friday Agreement of 1998. The reviews that we discuss in this section deal mostly with this issue as support to the civil power—an added role for the UK armed forces beyond the roles for which the forces are manned.

[5]   UK Ministry of Defence, *The United Kingdom Defence Programme: The Way Forward*, Cmnd. 8288, London: Her Majesty's Stationery Office, 1981.

[6]   UK Ministry of Defence, *The Falklands Campaign: The Lessons*, Cmnd. 8758, London: Her Majesty's Stationery Office, 1982.

[7]   UK Ministry of Defence, *The Falklands Campaign*, para. 313.

[8]   House of Commons, "Defence (Options for Change)," House of Commons Debate, July 25, 1990, *Parliamentary Debates*, Commons, 5th ser., Vol. 177, cols. 446–486.

[9]   House of Commons, "Defence (Options for Change)."

The First Gulf War (Operation Granby) (1990). Shortly after the announcement of the first stages of the *Options for Change* proposals, Iraq invaded Kuwait. The UK's response, conducted in conjunction with other coalition nations and led by the United States, was Operation Granby. A study of the lessons of this operation was reviewed by the House of Commons Defence Committee, which identified two areas of concern: (1) the vulnerability of ground troops and equipment to chemical and biological attack and the need for associated enhancements and (2) evidence that the operation had stretched logistical sustainability beyond a safe level.[10]

The Rifkind Mini-Review (1993).[11] This review's heralded increase in British Army manpower was to be paid for by reductions to the RN and the RAF. The intent was to improve logistic capabilities. The review was criticized for not relating national interests and associated priorities to available defense resources.[12]

Defence Cost Studies: Front Line First (1994).[13] In recognition of the continuing reduction in planned defense expenditures, the government introduced the *Front Line First* proposals to rebalance the armed forces toward greater operational capability. To do this, the review proposed that the force structure be changed and the use of joint structures be increased to avoid single-service duplication. Within the Ministry of Defence (MOD), a more powerful Central Staff was formed, the new Permanent Joint HQ (PJHQ) was established at Northwood near London, and the individual staff colleges were consolidated into a single entity as a tri-service joint staff college. Procurement decisions (such as the determination to procure new amphibious

---

[10] House of Commons, Defence Committee, *Implementation of Lessons Learned from Operation Granby, 1993–94*, London: Her Majesty's Stationery Office, 1994, paras. 17–35, 89–192.

[11] UK Ministry of Defence, *Statement of Defence Estimates: Defending Our Future*, Cmnd. 2270, London: Her Majesty's Stationery Office, 1993.

[12] House of Commons, Defence Committee, *Ninth Report: Statement on the Defence Estimates 1993*, London: Her Majesty's Stationery Office, 1993.

[13] House of Commons, "Defence Debate," July 14, 1994, *Parliamentary Debates*, Commons, 5th ser., Vol. 246, col. 1169.

ships or buy Tomahawk missiles for launch from RN submarines) reinforced the government's intention to enhance the ability of the armed forces to undertake deployed operations.

The House of Commons Defence Committee noted elements of a pattern to the reviews conducted in the last 40 years or so. The first element of this pattern is inconsistency.[14] However, the committee noted that

> inconsistency is not necessarily inappropriate to a changing world: for example, *Options for Change* delivered what was widely expected to be a "peace dividend" from the end of the Cold War, and established a force structure that has proved fairly durable.

The committee also discussed negative trends:

> Manpower has been on a steady downward trajectory, as has defence spending as a proportion of GDP . . . with ever increasing demands for improved efficiency. . . . It has not always been evident . . . that the demands on the armed forces have been reduced proportionately to the defence budget.[15]

This discussion of the defense reviews reveals that the British armed forces adopted a joint approach to reduce duplication of effort (and thereby save resources) but that lessons learned from major operations were incorporated into changes to ensure that the remaining forces were operationally effective. The specific case in point is the formation of PJHQ. Three single-service HQ, any one of which had, in theory, the capacity to command a joint operation (as the RN HQ did during the Falklands Conflict in 1982), were amalgamated into the single PJHQ structure. Lessons from the Falklands and the 1991 Gulf War were incorporated into the formation of PJHQ and the way it would operate with the support of the single services. In particular, the latter operation highlighted the huge demands that a British HQ

---

[14] House of Commons, Defence Committee, *Eighth Report: The Strategic Defence Review*, London: Her Majesty's Stationery Office, 1998.

[15] House of Commons, Defence Committee, *Eighth Report*, para. 49.

would face in modern warfare. Having three such HQ was deemed both infeasible given the size of the British armed forces and too expensive. PJHQ became operational in 1996 when it took over operations around the world (such as enforcing no-fly zones over Iraq in Operations Northern Watch and Southern Watch).[16]

It was against this background that the 1998 SDR, initiated by the new government in 1997, was developed. Without the successes and changes of the previous reviews—particularly those that introduced joint concepts that were put into practice and found to work—the SDR may not have been able to extend the use of joint administration and operational structures. The conclusions of that review, reported as a white paper in 1998, underpin the current activities of the British armed forces, although a series of white papers have since refined the SDR. We describe these white papers in more detail in later sections.

Sources

The UK introduced a freedom-of-information act in 2000 to give the public greater access to all levels of official activity; the act came fully into force in 2005. Consequently, a great deal of official information about the British armed forces is available online, notably at such sites as the UK Parliament's official Web presence (which supplies debates, announcements, studies, and so on) and the MOD's official Web presence, both of which have excellent search engines. We drew on these sources to establish points of fact and, in the case of parliamentary debates or reports of the Defence Committee, to add color to aid understanding. British officials in the UK and at the British Embassy in Washington, D.C., were very helpful during our research, providing information as SMEs in their own right or helping us to interview experts from relevant organizations in the UK.

---

[16] The increase in operational tempo is described in Tim Youngs and Mark Oakes, *Iraq: Desert Fox and Policy Developments*, International Affairs and Defence Section, House of Commons Library, Research Paper 99/13, February 10, 1999. A more general description that includes references to the role of the PJHQ is covered in House of Commons, Defence Committee, *Third Report: Lessons of Iraq*, London: Her Majesty's Stationery Office, 2004.

Strategic Demands and Focus

The SDR was to be a foreign policy–led review of the defense needs of the UK. As the report by the Defence Committee on SDR pointed out, "[i]t is difficult to identify a single concise statement of the government's foreign policy."[17] The committee instead pointed to the mission statement of the Foreign and Commonwealth Office (FCO), an organization conducting roughly the equivalent of the external responsibilities of the U.S. State Department. Since the publication of the 1998 SDR, the UK has clarified its foreign policy (most recently, in a 2006 white paper[18]), and it appears that these clarifications are sufficiently broad that they do not affect the assumptions that guided formulation of the defense priorities in the SDR or in later white papers. In the sections that follow, we refer to the original SDR and highlight instances when subsequent papers either refined the underpinning reasoning behind the SDR or altered priorities for the UK's armed forces.

---

[17] House of Commons, Defence Committee, *Eighth Report*, para. 82.

[18] UK Foreign & Commonwealth Office, *Active Diplomacy for a Changing World: The UK's International Priorities*, London: Her Majesty's Stationery Office, 2006. The document lists the following priorities:

A. making the world safer from global terrorism and weapons of mass destruction

B. reducing the harm to the UK from international crime, including drug trafficking, people smuggling and money laundering

C. preventing and resolving conflict through a strong international system

D. building an effective and globally competitive EU in a secure neighbourhood

E. supporting the UK economy and business through an open and expanding global economy, science and innovation and secure energy supplies

F. promoting sustainable development and poverty reduction underpinned by human rights, democracy, good governance and protection of the environment

G. managing migration and combating illegal immigration

H. delivering high-quality support for British nationals abroad, in normal times and in crises

I. ensuring the security and good governance of the UK's Overseas Territories.

Strategic Imperatives and Priorities

The SDR and its supporting papers were published in 1998.[19] In the documents, the interests of the UK were aligned to the continued well-being of the EU, including the stability and security of the European continent. In this context, NATO was seen as "a collective political and military instrument"[20] that allowed continued engagement in and between Europe and the United States. At the same time, the history and global interests of the UK meant that the country had interests and responsibilities, including in the 13 Overseas Territories, beyond Europe.[21] Additionally, the government wished to leverage the UK's permanent membership in the UN Security Council into an ability to play a leading role internationally. In the absence of a direct threat to the British mainland, the SDR determined that "national security and prosperity thus depend on promoting international stability, freedom and economic development" and that the UK therefore had a responsibility to act as a "force for good in the world."[22] The work of the SDR was updated in 2002 by the *New Chapter* volume,[23] which reflected on the lessons for the UK after the September 11, 2001, terrorist attacks. A 2003 white paper, *Delivering Security in a Changing World*,[24] clarified parts of SDR in light of world events and the experiences of the UK in managing its post-SDR armed forces. The final update to the *Deliver-*

---

[19] UK Ministry of Defence, *The Strategic Defence Review*, Cmnd. 3999, London: Her Majesty's Stationery Office, July 1998.

[20] UK Ministry of Defence, *The Strategic Defence Review*, para. 18.

[21] Anguilla, British Antarctic Territory, Bermuda, British Indian Ocean Territory, British Virgin Islands, Cayman Islands, Falkland Islands, Gibraltar, Montserrat, St. Helena and Dependencies (Ascension Island and Tristan da Cunha), Turks and Caicos Islands, Pitcairn Island, South Georgia and South Sandwich Islands. The Sovereign Base Areas in Cyprus are also considered a territory, but they were not listed as such in SDR.

[22] UK Ministry of Defence, *The Strategic Defence Review*, para. 21.

[23] UK Ministry of Defence, *The Strategic Defence Review: A New Chapter*, Vol. 1, Cmnd. 5566, London: Her Majesty's Stationery Office, 2002.

[24] UK Ministry of Defence, *Delivering Security in a Changing World*, Vol. 1, Cmnd. 6041-I, London: Her Majesty's Stationery Office, December 2003; UK Ministry of Defence, *Delivering Security in a Changing World: Supporting Essays*, Vol. 2, Cmnd. 6041-II, London: Her Majesty's Stationery Office, December 2003.

ing Security series is the volume subtitled *Future Capabilities*.[25] We align in this work our description of *strategic imperatives* to the term *security priorities*. The latter term, first used in the SDR, has been updated by subsequent policy papers into the following single *defence aim*:

> To deliver security for the people of the United Kingdom and the Overseas Territories by defending them, including against terrorism, and to act as a force for good by strengthening international peace and security.[26]

From this strategic imperative flow the more-detailed military tasks of the British armed forces, which are equivalent to the strategic priorities of other countries. Note that the UK does not accord greater importance to any one of the following sets of military tasks over the others:

- **Standing strategic tasks.** This group of military tasks covers the strategic elements of British defense policy, including the nuclear deterrent and strategic intelligence gathering. It also encompasses the provision of specialized contracted services vital to the armed forces' effectiveness. The tasks in this area include strategic intelligence; nuclear deterrence; and hydrographic, geographic, and meteorological services.
- **Standing home commitments.** These tasks encompass protection of British sovereignty, security at home in support of other government departments, and maintaining the armed forces' public profile. The tasks in this area include military aid to civil authorities, military aid to the civil power in Northern Ireland, maintaining the integrity of British waters, maintaining the integrity of British airspace, and carrying out public duties and transporting important persons.
- **Standing overseas commitments.** These long-standing tasks describe obligations to the 13 Overseas Territories, the UK's commitment to international alliances and partners as a means

---

[25] UK Ministry of Defence, *Delivering Security in a Changing World: Future Capabilities*, Cmnd. 6269, London: Her Majesty's Stationery Office, July 2004.

[26] UK Ministry of Defence, *Delivering Security in a Changing World*, Vol. 1, p. 4.

of safeguarding British interests overseas, and the promotion of the UK's influence and support around the world. The tasks in this area include defense and security of the Overseas Territories, defense and security of the sovereign base areas, and defense diplomacy (including supporting key alliances and partnerships; conducting arms-control outreach and other confidence- and security-building measures; promoting British interests and influence through, for example, military advisory teams; conducting defense exports; and supporting counterdrug operations).

- **Contingent operations overseas.** These seven tasks define the range of contingent commitments that may demand a contribution from the UK's armed forces. The tasks are humanitarian assistance and disaster relief, evacuation of British citizens overseas, peacekeeping, peace enforcement, power projection, focused intervention, and deliberate intervention. Obviously, these tasks are by no means mutually exclusive: Indeed, an operation may transition from one task to another during its lifespan. For example, an operation may start as a peace-enforcement operation and then shift to peacekeeping once a level of stability has been achieved.[27]

Force elements are assigned against each task using military judgment and an assessment of each task's requirements. This process of force summation is supplemented by modeling against agreed strategic scenarios that are bounded by the scale of effort expected of the UK's armed forces. In the land component, for example, a small-scale operation is defined as approximately battalion sized (500–1,000 personnel); a medium-scale operation as brigade sized (3,500–5,000 personnel); and a large-scale operation as division sized (10,000–20,000 personnel).[28] *Delivering Security in a Changing World: Supporting Essays*

---

[27] UK Ministry of Defence, *Delivering Security in a Changing World: Supporting Essays.* Our list reproduces the military tasks as described in the document but makes minor changes for the sake of clarity.

[28] UK National Audit Office, *Ministry of Defence: Assessing and Reporting Military Readiness*, London: Her Majesty's Stationery Office, 2005.

describes the overall scales of effort that the British armed forces should prepare to undertake:

> [Without creating overstretch,] . . . an enduring Medium Scale peace support operation simultaneously with an enduring Small Scale peace support operation and a one-off Small Scale intervention operation. [With a rapid reconfiguration,] . . . the enduring Medium Scale peace support operation and a Small Scale peace support operation simultaneously with a limited duration Medium Scale intervention operation. [With time to prepare,] . . . a demanding one-off Large Scale operation while still maintaining a commitment to a simple Small Scale peace support operation.[29]

The scenarios, which are classified, encompass the range of potential medium- and large-scale operations. These scenarios and the modeling are also used to determine the readiness profile needed for the force elements. The SDR and subsequent papers describe the modeling process as force estimation, and, through this measure and the one determined through force summation, the force levels are determined. The combination of the force elements and a given level of readiness forms the basis of the MOD's funding to the single services. The services are then responsible for manning and training their personnel and maintaining their equipment at the required readiness levels. This arrangement is the basis of the relationship between the single services and the joint operational structure that employs them. We consider the relationships between the military tasks, the force elements, and readiness in a later section.

The Role of Out-of-Country Deployments in National Strategy

Whereas earlier defense reviews had gradually increased the expeditionary nature of the armed forces, the SDR declared: "In the post Cold War world, we must be prepared go to the crisis, rather than

---

[29] UK Ministry of Defence, *Delivering Security in a Changing World: Supporting Essays,* Essay 2.

have the crisis come to us."[30] Out-of-country deployments, indeed, out-of-NATO-area deployments, were to be, and have become, the focus of the UK's forces. Since 1998, these deployments have been many and varied, and they have fallen under the full range of military tasks. Operations in Sierra Leone, Afghanistan, and Iraq are illustrative of the UK's willingness to deploy operational forces:

- **Sierra Leone.** The history of the UK's military involvement in Sierra Leone comprises a series of operations in 2000 and continuing support with ongoing military training and wider restructuring assistance.[31] The signing of a peace treaty ended the civil war in 1999, and, in that year, the UK sent a RN ship, followed by light forces in 2000, to aid the faltering government. These forces were augmented by rapid-reaction forces of battalion strength to prepare for the evacuation of noncombatants as fighting resumed. Later that year, the UK sent in a military training force and additional forces of more than battalion strength to rebuild Sierra Leone's army. British Special Operations Forces were then needed to lead a mission to recover captured trainers. This operation falls under the following military tasks: evacuation of British citizens overseas, peacekeeping, peace enforcement, humanitarian assistance, and defense diplomacy.
- **Afghanistan.** Since 2001, British armed forces have been involved in Afghanistan. British forces joined those of the United States under Operation Enduring Freedom to fight the Taliban and establish a legal government in Afghanistan. More recently, the UK has established Provincial Reconstruction Teams as part of the UN-mandated, NATO-led International Security Assistance Force, with the number of British forces committed increasing to a sustained total of 7,500. This operation falls under the following

[30] UK Ministry of Defence, *The Strategic Defence Review*, para. 6.

[31] Operations Basilica, Palliser, and Barras. See Jonathon Riley, "The U.K. in Sierra Leone: A Post-Conflict Operation Success?" The Heritage Foundation, Heritage Lecture No. 958, August 10, 2006. For an unofficial list of units that participated, see Britains–SmallWars. com, "UK Forces Deployed in Sierra Leone," Web page, 2008.

military tasks: power projection, focused intervention, peacekeeping, humanitarian assistance, and defense diplomacy.[32]

- **Iraq.** The UK joined other coalition forces under U.S. leadership of Operation Iraqi Freedom. The peak British commitment of 46,000 personnel occurred in the early part of 2003; the number was reduced to 18,000 personnel by May 2003, to 5,500 by the end of 2007, and to 4,100 by the end of 2008. This operation falls under the following military tasks: deliberate intervention, peacekeeping, humanitarian assistance, and defense diplomacy.[33]

How the UK's Strategic Imperatives Compare with Those of the United States

Although the UK describes its foreign and defense policies in a very different way compared with the United States, the coincidence of common interests and priorities is demonstrated clearly by the depth of commitment of both countries to the same operations. This shared commitment is not just a result of the special relationship. As the SDR and subsequent papers make clear, the UK, like the United States, sees itself as a force for good in the world, an upholder of international institutions and a collaborator in the maintenance of security and stability. These documents also highlight the larger context within which the UK executes its military tasks. The humanitarian assistance task is a good example of how the UK sees its role:

> The British are, by instinct, an internationalist people. We believe that as well as defending our rights, we should discharge our responsibilities in the world. We do not want to stand idly by and watch humanitarian disasters or the aggression of dictators

---

[32] There is significant material about the UK's involvement in Afghanistan. A good starting point is UK Ministry of Defence, "Operations in Afghanistan: British Forces," Web page, undated. For political commentary, there are many Defence Committee reports, the most comprehensive and up-to-date of which is, as of this writing, House of Commons, Defence Committee, *Thirteenth Report: UK Operations in Afghanistan*, London: Her Majesty's Stationery Office, 2007.

[33] House of Commons, Defence Committee, *First Report: UK Land Operations in Iraq 2007*, London: Her Majesty's Stationery Office, 2007; UK Ministry of Defence, "Operations in Iraq: Facts and Figures," Web page, undated.

go unchecked. We want to give a lead, we want to be a force for good.[34]

At the same time, there is a commitment in the UK to "ask our forces to fight [and] be sure they will win." [35]

The UK does diverge in some ways from the United States, however, most noticeably in its geographic interests. It is only recently that U.S. forces have returned to Africa (as part of the establishment of U.S. Africa Command and the Africa Partnership Station, among other initiatives).[36] As the Sierra Leone example shows, the UK and its armed forces have remained engaged in Africa. It is also difficult to see how the UK's interests in the Pacific align with those of the United States. Any British involvement in this region would most likely arise either because certain Pacific countries are part of the Commonwealth of Nations[37] or from the UK's participation in the Five Power Defence Arrangements.[38] For example, the UK participated in the UN-mandated International Force for East Timor, which undertook peacekeeping and humanitarian operations in East Timor in 1999–2000. The limitations on the UK's military involvement in this region most likely stem from financial constraints rather than from an unwillingness to engage in events that would trigger involvement closer to home. It should be noted,

---

[34] UK Ministry of Defence, *The Strategic Defence Review*, para. 19.

[35] UK Ministry of Defence, *The Strategic Defence Review*, para. 7.

[36] In making this comparison, we are differentiating between substantive involvement, which might be termed engagement, and transitory activities, such as emergency evacuations of personnel, reaction to attacks on U.S. embassies, and strikes against terrorists. U.S. engagement in Somalia (1993–1995) might be fairly described as the last period of engagement. See Lauren Ploch, *Africa Command: U.S. Strategic Interests and the Role of the U.S. Military in Africa*, Washington, D.C.: Congressional Research Service, Report RL34003, 2007.

[37] The Commonwealth of Nations comprises 53 sovereign nations bound by historical ties to the UK. With the exception of Mozambique, all 53 nations are former colonies of the British Empire.

[38] A South East Asia–centered series of bilateral agreements between the UK, Australia, New Zealand, Malaysia, and Singapore.

however, that, in the SDR and all subsequent white papers, the UK has accepted that global military activity may be necessary.

In broad terms, there is perhaps little of substance in the way of differences between the strategic imperatives of the United States and the UK. Both countries

- have nuclear deterrents
- provide for the protection of their nations (although the UK states that it sees no immediate threat and treats terrorism as a criminal act)
- support NATO and other international alliances fully
- are prepared to use military force overseas, although the UK acknowledges that, in the case of major or large operations, it will enter a coalition under the lead of the United States
- pursue policies characterized as a force for good.

Furthermore, the UK's approach to operations in Iraq and Afghanistan demonstrates a willingness to undertake the most demanding type of operations, sustain (relatively) large force levels for very prolonged periods, and share the U.S. commitment to both countries.

The Defense Establishment

The Components

The single services retain their traditions and ethos. Some joint units and facilities do exist; one example is the Joint Force Harrier, which provides RN and RAF Harriers for operations, and another is the Joint Command and Staff College, which is the UK's only officer-staff training venue. These joint organizations are considered to be more efficient than the single-service arrangements they replaced. The forces of each service are either employed in established operations or held at specified readiness levels in the event of a crisis. Established operations (which are typically of long duration) can be relatively benign or can involve warfighting. They can range from providing a frigate or destroyer to serve as a guard ship off the west coast of Africa to supporting UN forces in Bosnia to participating in such coalition operations as the enforcement of the no-fly zones over Iraq. There are three main components of the British armed forces and two additional organizations of interest.

The RN. The RN is one of the largest navies in the world. (The U.S. Navy is the largest.) Its major units include three small aircraft carriers, two landing ships, a helicopter carrier, about 25 frigates and destroyers, about 16 Mine Countermeasures Vessels, four ballistic submarines (which carry the nuclear deterrent), and nine attack submarines. The RN is supported by the **Royal Fleet Auxiliary,** an organization similar to the U.S. Naval Support Service, which has a variety of 17 ships that support deployed operations. The **Royal Marines** are also considered part of the RN. One Commando Brigade (3 Commando Brigade) commands three Commandos: 40 Commando, 42 Commando, and 45 Commando. Since April 2008, the Royal Marines have also commanded an attached infantry battalion. In addition, under the Royal Marines' command are specialist commando logistic and engineer units. The RN's air assets, maritime helicopters, commando helicopters, and Ground Attack Harrier pilots are part of the **Fleet Air Arm,** although the Harrier aircraft belonged originally to the RAF; the pilots and aircraft are also part of Joint Force Harrier, which falls under the RAF chain of command.

The British Army. There are two fighting (i.e., deployable) divisions that command the bulk of the forces in the Regular Army: 1st Division, which commands the 4th Mechanised, 7th Armoured, and 20th Armoured brigades; and 3rd Division, which commands 1 Mechanised, 12 Mechanised, 19 Light, and 52 Infantry brigades. There are four other divisions that fulfill regional command responsibilities, mostly for the Territorial Army units in their areas. In addition, there is 16 Air Assault Brigade, which, although under the command of the Joint Helicopter Command, can join the 1st or 3rd Divisions in an air-maneuver role.

The RAF. There are three RAF groups: 1 Group generates combat air power using attack and strike aircraft and combat-support helicopters; 2 Group focuses on air support to operations (including transport; air-to-air refueling; and intelligence, surveillance, targeting, and reconnaissance); 3 Group is responsible for recruiting and training RAF personnel. Deployed operations are conducted within Expeditionary Air Wings that are formed around the command structures of major British bases.

The British armed forces are now so intimately linked that, except for the most simple of deployments, all operations are approached from a joint perspective. This adds an important fourth component to the list, joint forces.

Joint Forces. The Chief of the Defence Staff (CDS, discussed in more detail in a later section) draws on his joint staff—in particular, the Commitments staff—to plan and direct operations. The Commander Joint Operations (CJO) heads the PJHQ and is responsible for campaign planning and the execution of directed operations. Force elements (made available by the single services because of their readiness status) are assigned to the CJO for any given operation. Embedded in the PJHQ is the JFHQ staff, who undertake command reconnaissance functions and can form the nucleus of the staff of a deployed joint task-force commander (JTFC). The JTFC is selected from a pool of one- and two-star flag officers (and their staffs) who have been trained by one of the single services. Logistics are a joint responsibility, are tied to the procurement process in the overarching Defence Equipment and Support organization, and are coordinated by the PJHQ to support operations. Routinely, the forces with the highest level of readiness (excluding special-operations forces, described in the next section) are assigned to the Joint Rapid Reaction Force (JRRF), which consists of units from each of the services that are configured as needed for emerging operations of medium scale or smaller. The Development, Concepts and Doctrine Centre provides the long-term vision used to develop the British forces' methods of operation. The center was established as a joint organization to ensure convergence of single-service doctrine and concepts.

Special Operations Forces. The UK releases very little official information about this fifth group. It mentions these forces in *Delivering Security in a Changing World: Future Capabilities* only to say that they will be enhanced.[39]

---

[39] UK Ministry of Defence, *Delivering Security in a Changing World: Future Capabilities*, para. 2.4.

Human Resources

The RN, the British Army, and the RAF are volunteer forces. Service personnel make a commitment to serve for specified periods, and they leave by either retiring or at a midcareer break point when they automatically become part of the Regular Reserve and may be recalled in emergencies. Each service may also draw on one of the following reserve forces:[40] the Royal Navy Reserve, the Royal Marine Reserve, the Territorial Army, and the Royal Auxiliary Air Force. The demands placed on the regular forces since 2000—particularly the British Army and certain specialist categories in the other two services—have resulted in increasing use of reserve personnel. Indeed, when the British government authorized that reservists be called up for Operation Iraqi Freedom, that was the first time such an authorization had been made since 1956.

Statutory Considerations

The British armed forces are under the political control of the prime minister and his or her secretary of state for defense. The government, which comprises the prime minister and the secretaries of the various ministries (who together form the cabinet), is supported by a neutral, career civil service. For example, the secretary of state is supported by the MOD, whose administrative functions are performed by civil servants and military officers. During operations and predeployment periods, civil servants provide ministers with political advice, and military officers provide military advice. The reality in the MOD is that military officers and civil servants work with and for each other to provide the ministers, and, ultimately the secretary of state and the prime minister, with the best advice possible. Much of the MOD is joint, although the single services retain their own staffs (military and civil servant) to support the following senior officers of each service: the RN's first sea lord, the British Army's chief of the general staff, and the RAF's chief of the air staff. The senior military officer is the CDS. The commander-in-chief of the UK's armed forces is Her Majesty the Queen, although this

---

[40] More information about the UK reserve forces is available in Directorate of Reserve Forces and Cadets, *Future Use of the UK's Reserve Forces*, February 7, 2005.

is mostly a ceremonial role. In practice, during operations, the prime minister, advised in a way that he or she determines,[41] instructs the CDS, usually via the secretary of defense, to execute an operation. In general, this instruction is accomplished by prime ministerial approval of a note from the secretary of state. The service chiefs do not have any operational role and are there to advise the ministers and the CDS in accordance with their own services' perspectives.

The legal advice for any intended or ongoing operation originates in the civil-military joint staff. These personnel discuss difficult matters with the appropriate civil servants in other government departments and, when necessary, seek professional legal advice from the staff of the attorney general. Whether to proceed with any particular course of action is a political decision made after the best legal and military advice and other factors have been taken into account. ROE are determined during this legal-review and political-decisionmaking process and are then reviewed throughout the operation. Changes to ROE require political approval.

Another key function of the MOD prior to and during an operation is negotiating with the Treasury to acquire any additional funding necessary to undertake the anticipated activities. Recall that the SDR determined what force elements need to be available (and upon what level of notice) to meet the UK's defense requirements; this determination is the basis of the UK's annual defense budget. The UK's armed forces are therefore funded to prepare for war or lesser operations, and they need additional funding, which must be preapproved, to start committing resources.

Finally, the MOD is where all military-related press matters are staffed and handled. Although most of the department's important announcements require political approval, the MOD is expected to issue announcements and develop media strategies that support national interests, not the interests of party politics.

---

[41] Some prime ministers have involved their cabinet in decisionmaking or have engaged in debate in the House of Commons to determine or justify an intended course of action (which may also involve a motion, debate, and a vote). Others have used a few members of the cabinet to form a smaller "war cabinet."

An earlier section in this chapter describes the British sense of right and wrong and how this worldview has been used to formulate aspects of the UK's defense policy. In this section, we describe the civil-military structure that operates within the country's government. A further aspect that affects political-military decisionmaking is the UK's parliamentary democracy, a system in which the decisions of the executive are placed under almost immediate, often adversarial scrutiny. Consequently, MOD staff and the ministers they serve strive to ensure that any military course of action, whether or not it involves combat operations, is both morally just and legal. In this regard, the decisions of the UN Security Council, the NATO Council of Ministers, and any of the many other internationally recognized competent bodies[42] become paramount in the minds of those who authorize the use of the UK's armed forces. It is also within this environment that the UK establishes and applies ROE.

Recruiting and Retention Considerations

Like most of the countries examined in this monograph, the UK is experiencing some difficulty in recruiting and retaining enough armed-forces personnel to meet operational demands. As of January 2008, the British armed forces employed 5,520 fewer personnel than authorized, a shortfall of 3.1 percent. As the information presented in Table 4.2 shows, however, the overall shortfall figures mask a steady demand for the British Army and reductions for the RN and RAF.

The best measure is probably that of the British Army, where the requirement has been constant for almost two years at 101,800 and the deficit has increased from 1,180 to 3,290.

The British armed forces face many of the challenges confronting the other countries we examined, challenges that make the military considerably less attractive as a career to the target population. The frequent separation and dislocation that are inevitably part of a military career increasingly deter prospective service members. More importantly, they deter their spouses, who would prefer neither to have their careers disrupted nor to regularly become single parents during

---

[42] Of particular relevance here are the European and International Courts of Justice.

Table 4.2
British Services' Full-Time Personnel Requirements, Strengths, and Surpluses/Deficits

	April 1, 2004	April 1, 2005	April 1, 2006	April 1, 2007	January 1, 2008
All services	195,340	191,090	185,920	183,610	180,430
Total requirement	195,340	191,090	185,920	183,610	180,430
Total strength	190,190	188,050	183,180	177,820	174,910
Surplus/deficit	−5,150	−3,040	−2,750	−5,790	−5,520
RN					
Total requirement	38,720	38,190	36,830	36,800	36,470
Total strength	37,510	36,400	35,620	34,920	35,200
Surplus/deficit	−1,210	−1,790	−1,220	−1,880	−1,280
British Army					
Total requirement	106,730	104,170	101,800	101,800	101,800
Total strength	103,560	102,440	100,620	99,350	98,510
Surplus/deficit	−3,170	−1,730	−1,180	−2,450	−3,290
RAF					
Total requirement	49,890	48,730	47,290	45,020	42,160
Total strength	49,120	49,210	46,940	43,550	41,210
Surplus/deficit	−770	480	−350	−1,460	−940

SOURCE: Adapted from UK Defence Analytical Services Agency, *UK Armed Forces Quarterly Manning Report*, TSP 4, London, 2008.

deployments.[43] The current security environment, with its high operational tempo, exacerbates these tendencies. The UK's National Audit Office, the British equivalent of the U.S. GAO, also attributed some of the recruiting shortfall to high operational tempo and the concomitant increase in workload and disruption to family life.[44] In recognition of such concerns, the MOD undertook in 2007 a review of its overarching manpower policies. Some details about intended changes to the policies have already emerged. For example, retention bonuses paid to enlisted personnel after four years of service, with a consequential commitment to continued service, are to be trebled from $10,000 to $30,000.[45]

As described in this chapter's introduction, the British armed forces have been under operational pressure for much of the last 20 years. In recognition of this and in an effort to highlight the importance of individual service members, the services developed guidelines (called *Harmony Guidelines*) for operational deployments and operational tempo. Although each service's interpretation of the guidelines differs somewhat, the guidelines are intended to provide some assurance of predictability in an environment of enduring operational commitments. For example, a soldier in the British Army can expect to spend six of every 30 months deployed to one contingency or another. Once again, the demands of Iraq and Afghanistan have made maintaining predictability somewhat difficult.[46] The other services apply the guidelines in a very similar way.

---

[43] The deterrent effects of the military lifestyle are described in Hans Pung, Laurence Smallman, Tom Ling, Michael Hallsworth, and Samir Puri, *Remuneration and Its Motivation of Service Personnel: Focus Group Investigation and Analysis*, Santa Monica, Calif.: RAND Corporation, DB-549-MOD, 2007.

[44] UK National Audit Office, *Ministry of Defence: Recruitment and Retention in the Armed Forces*, London: Her Majesty's Stationery Office, 2006, pp. 16–20.

[45] UK Ministry of Defence, "New Measures to Reward and Retain Forces Personnel," Web page, March 19, 2008.

[46] UK National Audit Office, *Ministry of Defence: Recruitment and Retention in the Armed Forces*, pp. 22–23

Priorities: The Mission Set and the Range-of-Operations Focus

The SDR recognized that maintaining high readiness demands greater resources and that careful management of personnel and materiel is central to the sustainability of the force structure. This point was reinforced by the National Audit Office in its 2005 assessment of the MOD's readiness system.[47] The MOD's approach to readiness is shown in Figure 4.1.

There are four phases between the first indication of a crisis and the point at which full in-theater operational capability is needed:

- **Decision time.** After the political-decision process, the Commitments staff of the MOD writes and issues (with appropriate approval) a CDS planning directive to the CJO. This directive attempts to describe the situation and the desired end-state;

Figure 4.1
The MOD Approach to Readiness and Warning Time

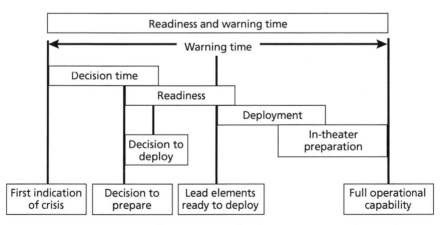

SOURCE: UK National Audit Office, *Ministry of Defence: Assessing and Reporting Military Readiness*, p. 8.
NOTE: Readiness is one of the four elements that constitute warning time. The others are decision time, deployment, and in-theater preparation.
RAND *MG836-4.1*

---

[47] UK National Audit Office, *Ministry of Defence: Assessing and Reporting Military Readiness*.

bound the operational area by time, space, and any other considerations; and provide any other available information. The MOD joint staff confers with other government departments and other nations to determine what role, if any, they will have in the operation. The PJHQ begins to develop an estimate and may push to deploy its JFHQ to aid both in this process and in the formulation of the campaign plan. The CJO responds to the planning directive with a plan that includes an estimate of forces required, resources required, and so on. This process may be iterated until the plan is acceptable and the decision to initiate the operation is reached. At this point, the CDS issues a directive for the CJO to execute the plan.

- **Readiness.** The potential size and nature of the operation in relation to the force elements at their level of readiness determine the speed with which the selected units reach the point of deployment.
- **Deployment.** The JRRF includes the necessary force-projection elements at the same level of readiness as the combat-force elements. This period also includes the time taken to arrive in theater.
- **In-theater preparation.** The nature of the operation and the environmental conditions determine the length of this phase.

The readiness categories and associated descriptions are shown in Table 4.3.

The MOD maintains an agreed readiness profile for its force elements, and this profile is related both to the force structure and to the resources assigned to the department. The scale-of-effort assumptions, together with other assumptions known as the *Defence Planning Assumptions*, guide the modeling of representative scenarios to determine the right balance of these variables. The single services and joint structures are then resourced to provide the force elements at the determined readiness level. Essentially, the single services train and staff their elements (which are, in simple terms, ships, battalions, and aircraft squadrons) and manage the availability of equipment in conjunction with the Defence Equipment and Support organization.

Table 4.3
British Readiness Levels

Readiness Category	Description of Force-Element Status
R0: immediate readiness	Ready to deploy; appropriately manned, equipped, and supported
R1: extremely high readiness	Ready at 2 days' notice
R2: very high readiness	Ready at 5 days' notice
R3: very high readiness	Ready at 10 days' notice
R4: high readiness	Ready at 20 days' notice
R5: high readiness	Ready at 30 days' notice
R6: medium readiness	Ready at 40 days' notice
R7: medium readiness	Ready at 60 days' notice
R8: medium readiness	Ready at 90 days' notice
R9: low readiness	Ready at 180 days' notice
R10: very low readiness	Ready at 365 days' notice
R11: very low readiness	Ready at more than 365 days' notice

SOURCE: UK National Audit Office, *Ministry of Defence: Assessing and Reporting Military Readiness*, p. 9.

The Range of Operations as Reflected in Preparations and Focus

The British armed forces are able to undertake a range of operations at relatively short notice. The JRRF gives British commanders the ability to tailor a force up to brigade size from each of the services and commence moving that force to the operational theater with 30 days' notice. Smaller elements (those up to battalion size) are held at an even higher level of readiness (i.e., five days or less) and can deploy in advance of the larger force when the situation allows or when operational requirements demand. There are two echelons of forces, as shown in Figure 4.2. The notice to deploy includes information about the relevant logistic assets; combat-support assets; and command, control, communications, and intelligence assets.

Force elements serving on peacetime deployment can be diverted to a higher-priority emerging operation, but this occurs only when the

Figure 4.2
JRRF Echelon Readiness

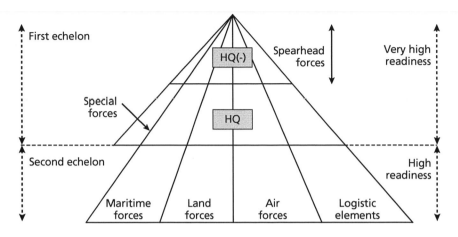

SOURCE: UK Ministry of Defence, Joint Warfare Publication 3-00, *Joint Operations Execution*, 2nd ed., London: Her Majesty's Stationery Office, March 2004.
NOTE: (-) indicates less than one full battalion, brigade, or other group.
RAND *MG836-4.2*

forces in the JRRF that are held ready for that purpose are deployed. Such a decision would likely only be made when the force elements serving on peacetime deployment possessed capabilities needed for the crisis operation.

Typically, the following units constitute the JRRF:

- first-echelon "spearhead forces" that are kept at a very high notice for deployment for operations (two days or less) and include
  - special forces
  - an attack submarine, surface warships, and a support ship
  - a spearhead battle group that is based on a light-infantry battalion or a commando group and is drawn from 3 Commando Brigade, 3rd Division's "ready brigade," or 16 Air Assault Brigade
  - a mix of combat aircraft, combat-support aircraft, helicopters, and supporting tactical air transport and air-to-air refueling aircraft

- a first-echelon balance of forces that includes
  - additional special forces
  - shipping to generate a maritime task group centered on an aircraft carrier and a helicopter assault ship; if necessary, amphibious shipping to support the lead commando battle group is supplied
  - lead battle groups, including a lead commando battle group, a lead air-assault battle group, a lead armored battle group, and combat-support and logistic-support groups
  - a range of air assets
- a second echelon that includes
  - additional maritime units to form a second or larger maritime task group, including amphibious shipping to support 3 Commando Brigade
  - a choice of brigades from 3 Commando Brigade; 16 Air Assault Brigade; and armored, mechanized, or infantry "ready" brigades from 1st Division and 3rd Division
  - substantial additional air assets.[48]

When the SDR was written, the JRRF was intended to include about 20 major warships, 22 other vessels (i.e., mine-warfare and support ships), four ground-force brigades, about 110 aircraft, and more than 160 other aircraft. Subsequent white papers in the *Delivering Security* series noted that the UK's experience in Iraq after 2003, combined with the frequency of other small- and medium-scale operations, led the country to review the JRRF, and they also noted that the ongoing high level of commitments had affected the second echelon in particular.[49] In practical terms, the JRRF commitment and "normal" peacetime enduring operations account for the majority of the operational forces of the British armed forces. The increased demands associated with supporting operations in Iraq and Afghanistan have made

---

[48] UK Ministry of Defence, *The Strategic Defence Review: Supporting Essays*, Cmnd. 3999, London: Her Majesty's Stationery Office, 1998, Essay 8.

[49] UK Ministry of Defence, *MOD Annual Report and Accounts 2006–7*, London: Her Majesty's Stationery Office, July 2007, para. 53.

it impossible for the single services to maintain the full range of other enduring operations and meet their commitments to the JRRF. This is particularly true in the British Army's case. Even so, the services have been able to maintain first-echelon forces, although the potential requirement to conduct further operations in the Balkans in early 2008 raised the prospect that even these first-echelon forces would be committed to current operations.[50]

An Assessment of Component Roles Against Mission Sets

The mission of the components of the British armed forces is to recruit and train the men and women they need to man their force elements— the ships, battalions, and squadrons or larger formations—then train those force elements to a predetermined level that will allow them to reach a specified level of readiness. It is also the responsibility of each single service to manage the allocation of its force elements to the readiness profile demanded of it. This involves planning maintenance; assigning force elements to training centers and exercises; monitoring harmony requirements; and assigning units to enduring tasks, which may require deployment, or to the JRRF.

Training Regimes

In this section, we consider how the single services train their force elements to the specific readiness levels and, when needed, provide further training in preparation for deployment. The RN and RAF approaches to training and manning are very similar to those of the U.S. Navy and the U.S. Air Force. In the case of the two countries' armies, however, there are greater differences. Thus, the major focus of this section is on assessing in more detail the British Army's approach to training and manning and the ways in which the service has reacted to the require-

---

[50] See, for example, Sean Rayment, "UK's Last 1,000 Soldiers Rushed Out to Balkans," *Daily Telegraph* (London), February 17, 2008.

ments associated with both preparing force elements for readiness and then generating these elements for current deployed operations.

The Royal Navy

The RN approaches the training of its ships much as the U.S. Navy does. All training is the responsibility of the Flag Officer Sea Training (FOST) organization.[51] As a ship emerges from an extended dockyard maintenance period, it progresses through a series of safety and readiness (SAR) inspections. The duration of these inspections depends on the progress of the work package in the dockyard and the complexity and number of new systems fitted to the ship. The SAR process is broadly similar to the U.S. Navy's unit-level-training readiness-assessment process. The following inspections constitute the SAR series:

- **SAR 1: safe to move on board.** This check, which occurs when the ship is still in dockyard hands, determines whether the work package is sufficiently complete such that it is safe for the ship's company to move back on board. In addition to habitability issues, the inspection covers all safety-related systems, such as the fire-fighting and damage-control equipment.
- **SAR 2: safe to go to sea.** As the dockyard work nears an end, a series of harbor-acceptance trials is undertaken to discover the state of key machinery and systems and to accept compartments and systems from the dockyard. The completion of this process is marked by other checks of, for example, personnel qualifications and individual competencies to determine that the ship can proceed to sea safely. The ship remains within adjacent exercise areas during the next phase.
- **SAR 3: safe at sea.** With the ship at sea or anchor, the ship's company practices emergency drills, including recovery of a man overboard, fire fighting, damage control, and navigation. Subsequent sea-acceptance trials of all systems are undertaken to con-

---

[51] A description of FOST's responsibilities and activities can be found in Royal Navy, "Flag Officer Sea Training," Web page, undated.

firm that they are operating correctly. For example, main engines are put through a series of speed and casualty tests.

After these checks, the ship is ready for more-demanding training. This training, Tier 1 Operational Sea Training (OST), is undertaken directly under the control of the FOST. There are three types of Tier 1 OST. Basic training is for ships emerging from dockyards and the SAR process. Directed Continuation Training (DCT) is tailored to the specific needs of ships about to deploy. Requested Continuation Training (RCT) can be requested by a commanding officer to sustain the operational capabilities of his or her ship. The early phases of Basic OST are comparable to the U.S. Navy's tailored ship training. DCT, the last phase of Basic OST, and Tier 2 OST are comparable to the integrated training events, such as joint task-force exercises (JTFXs), undertaken by U.S. Navy ships. RCT and DCT are also sustainment training activities. What a ship has to do is determined by the Mission Task List (Maritime), which is based on the U.S. Navy's Mission-Essential Task List (METL).

There are some differences between the U.S. Navy and RN systems, however. These differences stem from unlike expectations about expeditionary joint operations and from differing requirements to undertake joint training to achieve certain readiness levels. Looking more closely at OST organization and the different tiers exposes these differences:

- **OST organization.** The FOST trains all the ships and submarines of the RN and routinely trains most of the ships and submarines of many of the northern-European NATO nations. Other NATO and European nations send their ships to the FOST less regularly. Some nations pay for their training by providing the UK with training assets (notably, diesel submarines). Germany and the Netherlands have permanent liaison training officers on the FOST staff, and the French Navy has become more involved in FOST training since the late 1990s. FOST staff are considered to be some of the best experts in their fields of specialization in both the RN and the navies of other nations, and they train to high standards using NATO procedures. Normal training periods involve five or six ships from the RN or other nations and training assets (including foreign diesel submarines and aircraft) from both the UK and other nations.

- **Tier 1 Basic OST.** The training undertaken during this period includes every aspect of a running ship and involves every member of the ship's company. Training increases in intensity and complexity and culminates in a task-group exercise period during the final week. This allows the ships that had, until that point, been focusing on their individual capabilities to practice operations and procedures as part of a formed force.
- **Tier 2 OST.** The vehicle for this training period is a major exercise run as a joint maritime course (JMC). This exercise, called Exercise Neptune Warrior in recent years, is similar to a U.S. Navy joint force training exercise but generally involves more joint assets from a greater number of nations. These exercises can occur three times a year, although current commitments have reduced this to twice a year, and rotate through themes to match the participating assets. For the maritime units, the exercise theme is always of an expeditionary nature. For example, if an amphibious ready group with an embarked commando unit is involved, the exercise is biased in favor of littoral operations and amphibious landings. RAF and British Army units participate as either friendly or opposing forces, and most other European NATO nations send their own forces to train. A typical Exercise Neptune Warrior involves 20 ships, several submarines, and nearly 100 aircraft.
- **Mobile OST.** FOST mobile teams are available to provide tailored training to deployed ships or units activated to undertake an operation. In the case of ships, the FOST teams join the ships overseas or sail with them to the vicinity of the operational area to assist in continuation training or specific deployment training.

Like U.S. Navy ships, RN ships need to achieve a high level of capability irrespective of the nature of the operations that they may be required to undertake. Tailored deployment training is easily accommodated in a relatively short period before departure or during passage to the operational area. Throughout its operational training, the RN takes advantage of the UK's proximity to other NATO nations, thereby allowing its ships and submarines to easily integrate into multinational groups during coalition operations. The RN's routine process of training with RAF and British Army units during the JMC exercises underlines how embedded expeditionary joint operations are in the mindset of RN commanders.

C2 training and certification for the one- and two-star flag officers who could become JTFCs or maritime component commanders is undertaken during Tier 2 training, when staff are expected to not only run the exercise but also react to high-level control inputs that simulate interaction with the PJHQ. These staffs are further exercised during NATO or bilateral exercises, such as when British forces participate in U.S. JTFXs.

In much the same way that the U.S. Navy programs its ships and submarines so that they can come together to form carrier, expeditionary, or surface strike groups, the RN coordinates its force elements to maintain the required readiness profile and ensure that they are ready either to meet any enduring commitments or to deploy as a task group in emerging operations.

The Royal Air Force

The RAF has responded to lessons learned from recent operations with a greater focus on the deployed organizations that support and enable the delivery of force elements. The training of RAF force elements, which has remained relatively unchanged, follows the Mission Task List (Air). The force elements, principally the aircraft squadrons, are evaluated and certified using NATO's tactical-evaluation program.

The basis of RAF deployed operations is the Expeditionary Air Wing (EAW). There are nine peacetime EAWs formed on the existing main operating bases (MOBs) of the RAF in the UK. The key personnel and structures in the "home construct" are those of the base commander and his or her staff. The force elements at readiness (i.e., the squadrons at the base) are not formally part of the EAW, although they remain under the C2 of the base commander. Two additional key force elements support the base: Air Combat Support Units (ACSUs) and Air Combat Service Support Units (ACSSUs). Both types of unit provide training to the squadron force elements at both the MOBs and other bases in the UK.

When an EAW is activated, the base commander, his or her staff, and the MOB support functions deploy to an air point of departure or deployed operational base (DOB). At this point, force elements at readiness from either the original MOB or elsewhere (depending on operational requirements) become part of the EAW. Finally, dedicated ACSUs and ACSSUs join the EAW to provide the necessary training and support. The home and deployed constructs of an EAW are shown in Figure 4.3.

Deployed operational training of EAWs is coordinated by the Air Training Division staff of HQ Air Command. The levels of training for an EAW are shown in Figure 4.4. At the individual level, common core skills and individual readiness training are the responsibility of the MOB and are part of routine training.

The EAW concept was introduced by the RAF in 2006, and the exercises and procedures that will make it work are still evolving. Our

Figure 4.3
Home and Deployed RAF EAW Constructs

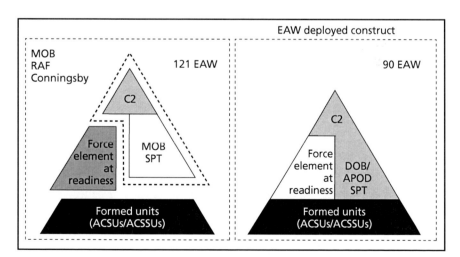

SOURCE: UK Ministry of Defence, Development, Concepts and Doctrine Centre, JDP 4-00, *Logistics for Joint Operations*, 3rd ed., April 2007.
RAND *MG836-4.3*

Figure 4.4
Progressive Levels of RAF EAW Training

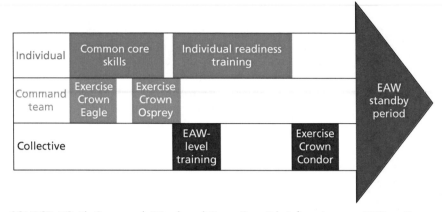

SOURCE: HQ Air Command, "Deployed Operations," briefing, January 2008, p. 3.
RAND *MG836-4.4*

description of training in this section is based on our interviews with RAF staff at HQ Air Command and their January 2008 briefing on deployed operational training.[52] Exercises Crown Eagle and Crown Osprey, named in Figure 4.4, are tabletop events designed to train the EAW command teams. Crown Eagle is a three-day basic exercise that is set in an unfamiliar country and requires the command team to address the factors involved in reconnaissance, planning, and establishment of a DOB. Crown Osprey is a two-day exercise that allows the command team to explore how to manage the risks involved in the sustainment and operation of a DOB.

Collective training is much more demanding. The first level, EAW-level training, is the responsibility of the base commanders (with assistance from external assets) and occurs annually. EAW-level training builds on individual skills training, which is already part of the training structure at a base, and refines these skills in a simulated expeditionary location. Exercise Crown Condor is a major exercise that involves the deployment of an EAW with the necessary force elements to establish a DOB. The exercise includes a three-day warfighting phase

---

[52] HQ Air Command, "Deployed Operations."

during which the EAW's interaction with a higher-level command is practiced through the involvement of Air Command staff.

The British Army

The British Army doctrinally recognizes and clearly articulates the distinction between force preparation (preparations for war in general) and force generation (preparations for "the" war or a specific operation). It maintains separate but integrated mechanisms for each process, combining the two processes effectively to prepare forces, especially at the battle-group level and below, for a wide variety of operational environments (see Figure 4.5). This approach parallels the funding requirement of force elements to readiness (force preparation) and the additional funding provided for current operations (force generation).

Units devote the bulk of their time and effort to force preparation oriented on traditional challenges; this is also referred to as *building the adaptive foundation*. They do so not only because the readiness profile calls for force elements with these skills but also because the British Army believes that these skills provide the basic, essential foundation

Figure 4.5
Training for Operations

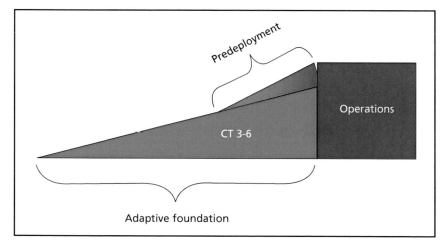

SOURCE: British Army, Land Warfare Centre, Operational Training and Advisory Group, "Operational Training and Advisory Group Brief," briefing, 2008.
RAND *MG836-4.5*

of all other military operations. Training at the CTCs generally forms the capstone of the force-preparation process.

The British Army then builds on this adaptive foundation with operationally specific training, which includes both predeployment training and in-theater training. JRRF forces with a very high level of readiness are more likely to conduct in-theater training, and the second-echelon forces would likely take advantage of both types of training. The high tempo of current operations and the nature of the UK's enduring commitments worldwide, including those in Afghanistan and Iraq, have led to the development of the Operational Training and Advisory Group (OPTAG), an organization that helps commanders prepare for specific operational environments. OPTAG focuses primarily on training unit leaders, who then become responsible for training their units. OPTAG ensures that the training provided by unit leaders has been effective by conducting a confirmatory exercise. Unlike the United States, where the distinction between force preparation and force generation has blurred and both models have become oriented almost exclusively on Iraq and Afghanistan, the UK (and OPTAG in particular) has maintained the capability to prepare forces for service in a variety of operational environments, including Iraq, Afghanistan, the Balkans and, to a diminishing extent, Northern Ireland. The British Army's Copehill Down urban-operations training facility also plays an important role in predeployment training.[53]

The British Army's Manning Strategy. It is difficult to analyze training policies in isolation from manning and education policy. The British Army's manpower policies emphasize seniority and cohesion. Leaders and those they lead serve in the same regiment for their entire careers or, at the very least, for as long as they serve at the battle-group level and below. Finally, British Army policy tends to emphasize service in foreign contexts and with foreign organizations to a significantly greater degree than does U.S. Army policy. The British Army's professional-development system follows approximately the same pattern as that of the U.S. Army, but British courses are often attended

---

[53] House of Commons, Select Committee on Defence, "Written Answers for 10 October 2007," October 10, 2007, *Parliamentary Debates*, Commons, 5th ser., Vol. 464, col. 636W.

by a significant proportion of foreign officers, a feature that increases British Army personnel's exposure to foreign cultures. Taken together, these factors create a different set of conditions for British officers and enlisted personnel as they learn their trade, a set of conditions that maximizes the value of training received and emphasizes adaptability to foreign cultural contexts.

On the whole, British officers are slightly older than their American counterparts, and they are considerably older than their American counterparts when the comparison considers officers with equivalent responsibilities. The average age of the British officer is 36; the average age of the American officer is 34. In the British Army, senior lieutenants (and, sometimes, captains), who have 1–5 years of experience, command platoons; majors, who have about 11 years of experience, command companies. In the U.S. Army, newly commissioned officers with no operational experience lead platoons; captains, who have about 6–8 years of experience, command companies.[54]

Professional military education occurs at roughly similar points in the careers of British Army and U.S. Army officers, but there is somewhat more of it, especially distance learning, in the British Army. To begin with, British Army officers enter the service with a very high degree of education. Over half have matriculated from private boarding schools, which offer education of very high quality.[55] Moreover, most have also graduated from university. The one-year course at the Royal Military Academy, Sandhurst, serves much the same function as the new U.S. Army Basic Officers Leaders Course, although it lasts longer and goes into greater depth. Prior to joining his or her regiment, a young officer also attends a branch-specific school equivalent to a U.S. Army basic

---

[54] The average age of the British officer is from UK Ministry of Defence, *UK Defence Statistics 2007*, London: Defence Analytical Services Agency, September 2007, Table 2.8. U.S. information is from U.S. Department of Defense, Office of the Under Secretary of Defense for Personnel and Readiness, *Population Representation in the Military Services, FY 2004*, May 2006, Chapter 4. Information about levels of command in the British Army is from British Army, *Officer Career Development*, July 1, 2003, Table 1.1.

[55] House of Commons, Committee of Public Accounts, *Recruitment and Retention in the Armed Forces: Thirty-Fourth Report of Session 2006–07*, London: Her Majesty's Stationery Office, 2007, p. 14.

course. For the next ten years or so, officers combine a series of shorter resident courses, guided self-directed study, and distance education to improve their professional qualifications. Shortly after entering the zone for promotion to major (after approximately ten years of service), officers attend the Initial Command and Staff Course (Land), which prepares them for company command and staff positions appropriate to their rank. Similarly, after entering the promotion zone for lieutenant colonel (after approximately 17 years of service), selected officers attend the Advanced Command and Staff Course, which is a prerequisite for command. Selected colonels and brigadiers also attend the Higher Command and Staff Course to prepare to assume operational and strategic responsibilities. The overall pattern of British Army officer development is that all officers attend courses throughout their careers and engage in continuous self-directed study.[56] Professional education also contributes significantly, though indirectly, to officers' cultural adaptability. As previously noted, a significant number of foreign students attend the UK's military schools, beginning with Sandhurst. Usually, two or three cadets out of every platoon are foreigners.[57] A similar proportion also attends higher-level military-education courses.

Adventurous training is an aspect of leader development that receives considerable emphasis in the UK yet has no analog in the United States. In purpose, adventurous training resembles French battle-hardening training, although the British version is less structured. In the UK, individuals and groups participating in this type of training engage in challenging outdoor pursuits, such as skiing and mountaineering, in order to develop leadership and character attributes held to be essential in wartime. Training groups are often ad hoc, and the training is not necessarily linked to military tasks.[58] Still, adventurous training does present individuals and teams with opportunities to confront challenging and unfamiliar tasks, and it may contribute to the development of adaptability.

---

[56] British Army, *Officer Career Development*.

[57] Royal Military Academy Sandhurst, "Overseas Cadets," Web page, undated.

[58] Most of our description of British predeployment training comes from discussions with British Army officers, Washington, D.C., summer and fall 2007.

The British Army's Overall Training Methodology. The British Army's training management begins with operational requirements. Like the armies of the United States and France, the British Army employs a cyclic readiness system. In the UK, this model is known as the *Force Operations and Readiness Mechanism* (FORM). FORM, depicted in Figure 4.6, is a logical, sequential mechanism that progressively increases unit proficiency through deployment.

The goal of the readiness cycles is to prevent a soldier from spending more than six of every 30 months deployed. Operational requirements in both Iraq and Afghanistan, however, combined with ongoing commitments in the Balkans and elsewhere, have placed significant

Figure 4.6
The Force Operations and Readiness Mechanism

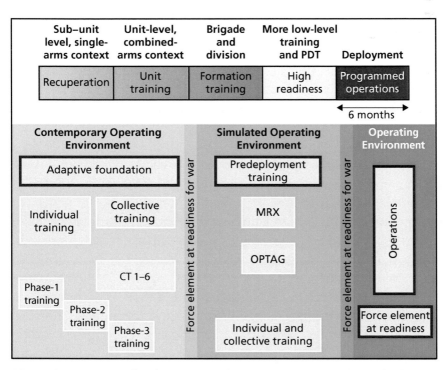

SOURCES: UK Ministry of Defence, HQ Land Forces, "Force Preparation and Generation," briefing, 2008; Rupert Pengelley, "Reality Check: Learning the Art of War in an Age of Diverse Threats," *Jane's International Defence Review*, November 1, 2006.

RAND *MG836-4.6*

pressure on that model. Nonetheless, this model implies that a British unit can expect to spend about one-fifth of its time deployed; the expectations for French and U.S. units are one-quarter and one-third to one-half, respectively. In the UK, this proportion therefore indicates a relatively greater requirement for effective training, rather than experience, to develop relevant individual and collective skills.

In forecasting available forces against operational requirements, HQ Land Command tentatively assigns forces against actual missions or designates them as a contingency force to be held in readiness. Commanders then develop training directives, which prescribe the specific military tasks that subordinate units must prepare to perform. Like the U.S. Army, the British Army derives these tasks from a doctrinal list. This list, the Mission Task List (Land), differs from its U.S. analog, the Army Universal Task List, in that it describes *what* is to be accomplished but does not address *how* the task is to be accomplished or under what conditions. Furthermore, unlike commanders in the U.S. Army, British commanders prescribe training tasks to subordinate commanders using a training directive. The distinction is subtle. Under U.S. doctrine, subordinates develop their METL, and superiors approve it. In the British system, superiors develop the training directive in consultation with subordinates.

After developing the training directive, commanders begin to develop their training programs to prepare their units for the specific operational environments in which they will be employed. Although individual commanders are not free to choose the tasks for which they train, they retain almost complete autonomy regarding when and under what conditions training is accomplished. Thus, commanders are responsible for integrating force preparation and force generation to deliver an effective unit.

Force Preparation: The Adaptive Foundation. The term *adaptive foundation* refers expressly to the British Army's force-preparation measures, which provide the basis from which forces adapt their military capabilities to a specific operational environment. The adaptive foundation emphasizes traditional individual and collective skills. British military officials state that this approach teaches soldiers and units the basic soldier skills and C2 processes that are essential in all operations.[59] The measures consist of a predictable sequence of individual and collective training events that are expected to prove useful in the various operational environments in which units may be employed.

Development of the adaptive foundation proceeds in stages that allow a battle group to gradually attain its full measure of proficiency. The five levels of collective training (CT) are

- CT 1: training up to the troop or platoon level
- CT 2: training at the sub–unit level (i.e., company or squadron)
- CT 3: training at the sub–unit level in a task-organized unit or combined-arms battle-group context
- CT 4: training at the task-organized unit or battle-group level in a combined-arms formation context
- CT 5: training in a brigade-sized formation context
- CT 6: training in a division-sized formation context.[60]

British divisions generally use one of the major training estates as a capstone exercise to train their brigades to full proficiency.[61]

CTCs. The British Army operates three major maneuver training centers (known as Army Training Estates) and several smaller ones. The major training estates are the Battle Group Training Unit (BGTU) at Salisbury Plain, which supports mechanized-infantry battle-group training; the training estate Sennelager, Germany; and the largest of the three, the British Army Training Unit in Suffield, Canada (BATUS), which supports armored and mechanized battle-group training. All three training centers can support training with direct-effects and area-weapon–effects systems similar to those employed at U.S. CTCs. Dedicated opposing-force units support training at BGTU and BATUS. The British Army's Land Warfare Centre also has an exportable training capability and has deployed a simulation-supported training package to Poland to support brigade exercises. Rotations at the

---

[59] Discussions with British Army officers, Washington, D.C., summer and fall 2007.

[60] Pengelley, "Reality Check."

[61] Pengelley, "Reality Check."

major training estates usually complete the process of establishing the adaptive foundation. Finally, the British Army operates light-infantry training centers in Belize and Kenya, but these do not enjoy the robust resourcing received by BGTU or BATUS.

The British Army also operates three Command and Staff Trainers (CASTs) at fixed sites in the UK and Germany. CASTs support both force preparation and force generation and emphasize the staff planning process rather than the particulars of the scenario. CASTs train staffs at the battle-group level and above, using the Advanced Battlefield Computer Simulations (ABACUS). At present, the ABACUS simulation can only support traditional combat operations.[62]

Finally, armored and mechanized forces can train using one of two Combined Arms Tactical Trainers (CATTs), which are analogous to the U.S. Close Combat Tactical Trainer. CATTs, located in the UK and Germany, allow battle groups to conduct maneuver training at the battle-group level in a simulated environment. Like CASTs, CATTs focus mostly on traditional combat operations, although the British Army is moving to incorporate more-complex features—principally, urban terrain and noncombatants—into the simulation.[63]

Originally established to support training for high-intensity conflict, BGTU and BATUS have recently begun to expand their focus to the less-intense but more-complex operations that occur in the middle of the spectrum of conflict. BATUS, for example, has added a mock Arab village and tunnel complexes to its site. The two training estates are also developing an urban live-fire complex that will allow two companies to maneuver at once. Training tests not only how well commanders and units react to a given scenario but also how quickly and how well they can shift between scenarios. Nonetheless, a recent BATUS commander is very clear that his primary purpose is to maintain foundational skills in high-intensity, armored warfare.[64]

---

[62] Discussions with British Army officers, Washington, D.C., summer and fall 2007.

[63] Pengelley, "Reality Check," pp. 3-9–3-10.

[64] UK Ministry of Defence, "Exclusive: Back to the Future: Army Training Is Ahead of the Game in Canada," Web page, October 29, 2007.

British Army Deployment Training. The British Army's force-generation efforts emphasize training conducted by OPTAG, which prepares units and individuals for the specific operational environment (e.g., Iraq, Afghanistan, the Balkans, Northern Ireland) for which they are destined. To facilitate this process of ongoing adaptation from operational lessons learned, OPTAG maintains continuous surveillance of ongoing operations, including through frequent trips to theaters of operation. OPTAG does not actually conduct training, however; instead, it operates on a train-the-trainer principle. OPTAG provides theater-specific training to unit leaders, who then train their units. OPTAG assesses the success of this effort during a confirmatory exercise and, if necessary, conducts retraining at the request of the unit commander.[65]

The overall OPTAG process, depicted in Figure 4.7, begins with a reconnaissance trip to the theater of operations. This trip, which is analogous to a U.S. predeployment site survey, is conducted jointly by unit leaders and OPTAG personnel. The purpose of this reconnaissance is to identify what the unit will have to do to prepare itself for its upcoming mission. During or immediately after that trip, the unit commander and OPTAG establish training objectives to be supported by OPTAG and establish the schedule for OPTAG's training support.

Training starts with a brief to the entire unit on the particular characteristics of the area of operations, including the local population, the enemy, and the most-effective tactics being used by all belligerents (friendly, enemy, and otherwise) in the area of operations. OPTAG then delivers to unit leaders a period of intense training designed to enable those leaders to train their subordinates in the relevant aspects of the operational environment. This training is oriented primarily at the company level and below. The leaders then accomplish the theater-specific training for the deploying unit.

To ensure that the lessons have been effectively learned, the unit returns to OPTAG for a one-week confirmatory exercise that is focused mostly at the platoon level and below. This confirmatory exercise is not supported with either direct-fire-effects or area-fire-effects simula-

---

[65] Discussions with British Army officers, Washington, D.C., summer and fall 2007.

Figure 4.7
A Typical OPTAG Training Cycle

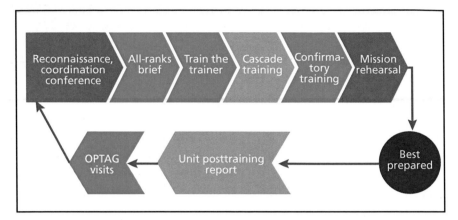

SOURCE: British Army, Land Warfare Centre, Operational Training and Advisory
Group, "Operational Training and Advisory Group Brief."
RAND *MG836-4.7*

tion, but it does portray essential elements of the operational environment with exceptional realism. Civil-disturbance training can get quite
rough, featuring pyrotechnics and frequent physical clashes. The aim
of these exercises is to ensure that soldiers and small units demonstrate
the knowledge and skills necessary to succeed in their operational environment. If they require additional training, they receive it.

It is significant, however, that OPTAG does not similarly assess
the performance of unit commanders or their staffs. This assessment
is conducted during the unit's mission-rehearsal exercise, which usually takes the form of a CAST exercise for the unit commander and
his or her staff.[66] In fact, at no time during this process does OPTAG
either grade the unit's performance or provide any sort of assessment to
higher ups within the British Army hierarchy.

An important follow-up component of the OPTAG process helps
the process maintain its relevance and vitality: About three months
after deployment, the deployed unit provides an assessment of the
value of the training. Although this assessment conforms to the British

---

[66] Discussions with British Army officers, Washington, D.C., summer and fall 2007.

Army's doctrine of training analysis, it is practiced more regularly and pays more attention to the training unit's perspective than is perhaps the case in other venues. In short, the unit grades the training center, not the reverse.

The British value adaptability and improvisation, but their training program clearly falls squarely within the category of ongoing adaptation based on lessons learned. There is little apparent effort devoted to forming and developing commander-leader teams or developing the skills and intuition necessary to operate in a strategic time frame, elements the IDA study identified as characterizing adaptability training.[67] Against the absence of apparent developmental efforts, however, must be set the effects of the British regimental system, which may achieve many of the same effects as the deliberate formation and training of commander-leader teams. The UK's efforts to inculcate adaptability focus on military education and developmental assignments, which expose individuals to a high degree of contact with people from other cultures and frequently immerse those individuals in those cultures. British officers have become quite used to serving on U.S. and other multinational staffs and to being the lone Westerner on UN peacekeeping missions.

This approach appears to be fairly effective at the battle-group level and below, where the British have placed their training emphasis. During Operation Iraqi Freedom, British soldiers and small units have, it appears, been quite effective in transitioning between combat and stability operations, and they have been able to shift back and forth between the two as necessary. British soldiers have demonstrated awareness of, and sensitivity to, the cultural context in which they operate in places as varied as Iraq, Afghanistan, Sierra Leone, and Northern Ireland.

Assessing the Effectiveness of the British Army's Training. The British Army's participation in Operation Iraqi Freedom provides the best basis on which to assess the value of British Army training models and methodologies. To begin with, this participation has involved most of the full range of military land operations. Second, the British experi-

---

[67] Tillson et al., *Learning to Adapt to Asymmetric Threats*, pp. 69–70.

ence in southern Iraq, especially Basra, required mastery of a complex and volatile political, military, and social context and required many sudden transitions across the range of military operations. Third, the involvement has lasted for some time, requiring the rotation of British forces and, consequently, preparing them for operations on a systematic and continuing basis. Fourth, this participation is better documented, and has received more analysis and attention, than any of the UK's other recent military operations.

The UK was a key member of the U.S.-led coalition from the outset and eventually contributed approximately 46,000 troops at the peak of its involvement. The UK did not, however, fully commit to participating in the invasion until planning was well under way, which limited its ability to influence those plans. This proved especially important with regard to the critical stabilization and reconstruction phases, which the British recognized as deficient from the outset of their involvement. The UK's land-force contribution consisted mainly of the HQ of the 1st Armoured Division, commanded by Major General Robin Brims. General Brims commanded two maneuver brigades, the 7th Armoured Brigade and the 16th Air Assault Brigade, with supporting elements.

General Brims's principal missions were to seize the southern Iraqi city of Basra and to open the port of Umm Qasr. In conducting these missions, he displayed considerable sophistication. He and his commanders were concerned that a direct assault on Basra, with attendant collateral damage and high casualty rates, would drive the city's residents into the arms of the Ba'athist regime. Instead, General Brims decided to induce regime collapse by isolating the city and launching targeted attacks against regime power centers, including the military and intelligence HQ and the Ba'athist HQ. Despite some initial anxiety about the pace of operations, the plan worked beautifully, and Basra fell to the 7th Armoured Brigade like a ripe peach on April 6, 2003.[68]

In general, British forces handled the transition to stability operations speedily and well. The most visible sign of this adept performance was the quick transition from full battle gear and armored vehicles to

---

[68] House of Commons, Defence Committee, *Third Report*, pp. 97–99.

berets and Land Rovers, which evinced a willingness to engage with the population rather than fire at it. There was more to the transition than a good attitude: British commanders reached out to local leaders and immediately initiated quick-impact projects to address local concerns about security, infrastructure, and the economy. As late as March 2005, the House of Commons Defence Committee contrasted the relative calm and stability in Basra with the chaos besetting U.S.-occupied Iraq, attributing the former situation to the professionalism and competence of British soldiers.[69]

By 2006, however, the situation in Basra had begun to deteriorate. In May 2006, Iraqi Prime Minister Nouri al-Maliki declared a month-long state of emergency in the city.[70] Rival Shi'a militias, including Moqtada al-Sadr's Mahdi Army, the Badr organization, and other, more-local gangs, began to contend for power and control over Basra and the rich resources of southern Iraq. Until that point, the British had trod very lightly with regard to the Shi'a militias, viewing them as an Iraqi problem for the Iraqis to work out. Nonetheless, British forces became the target of all factions because they both posed an obstacle to factional domination and formed a relatively legitimate target against which militias could demonstrate their military potency. The situation came into sharp relief in September 2005 when British troops stormed a Basra prison to free two British service members being held by militia forces.[71]

We disagree with Anthony Cordesman's suggestion that it was the British who "lost" Basra in 2005. To begin with, Basra's internecine warfare is fueled by forces well beyond southern Iraq. Second, the British lacked the manpower necessary to impose stability on Basra: Maintaining a brigade in Iraq and an equivalent (or greater) number of forces in Afghanistan has strained the British Army, which already suffers from a severe recruiting shortfall. Third, the mandate to support the national Iraqi government may simply have been unrealistic given

---

[69] House of Commons, *Iraq: An Initial Assessment of Post-Conflict Operations*, London: Her Majesty's Stationery Office, 2005, p. 4.

[70] House of Commons, Defence Committee, *Thirteenth Report*, pp. 6–8.

[71] "Kennedy's Fear of Iraq Civil War," BBC News, September 20, 2005.

the weakness of the Maliki government and the lack of political development within Iraq.[72]

Although the deterioration in Basra cannot be attributed to the conduct of British forces in the province, that deterioration makes it difficult to assess whether there are any lessons for the DoD to learn from British Army training methods for higher-level formations. Such an assessment is especially difficult given the degree to which British and U.S. methods are similar. According to a British officer who served with the brigade HQ in Basra, British HQ were always aware of who was doing what to whom, for which reasons, and to what effect.[73] Moreover, British forces have performed admirably at the battle-group level and below. From their seizure of Basra until their withdrawal from the city, they fought well and bravely when necessary and conducted themselves with admirable restraint and sensitivity. British attacks in the city were precise and limited to what was absolutely necessary, but they were also extremely effective. The fact that the forces were able to maintain this standard of performance and conduct over several rotations of British forces over almost five years speaks well for their preparations at that level and recommends emulation.

Comparison of British Army and U.S. Army Regimes. The key difference between British Army and U.S. Army training regimes is the explicit differentiation between force preparation and force generation. In the British model, the purpose of training for traditional military operations is as much to teach the fundamentals of soldiering as it is to prepare soldiers for their likely tasks. In this vein, units go to BATUS to learn how to operate as a unit, not because the British Army thinks it likely that they will face a mechanized enemy in the near future.

That explicit differentiation, along with the relatively lower operational demands placed on British forces, allows the British Army, which is much smaller than the U.S. Army, to prepare for a wider range of threats using the mechanism of OPTAG. In the current U.S. system,

---

[72] Anthony Cordesman, *The British Defeat in the South and the Uncertain Bush 'Strategy' in Iraq: 'Oil Spots,' 'Ink Blots,' 'White Space,' or Pointlessness?* Washington, D.C.: Center for Strategic and International Studies, 2007.

[73] Discussions with British Army officers, Washington, D.C., January 17, 2008.

CTC rotations prepare units for a specific operational environment. Because there are only three U.S. CTCs, all of which are oriented on Iraq and Afghanistan, using those CTCs to prepare for another operational environment would involve retooling the entire system. In contrast, OPTAG allows the British Army to prepare for a wide variety of operational environments that range from Northern Ireland to the Balkans to Iraq and Afghanistan.

Training for Senior Officers. Like most of the countries we researched, the UK has a robust officer-education system. Senior-officer training does, however, include a course focused on continuing education past that available in U.S. senior service colleges (war colleges). This course, the Higher Command and Staff Course, is "a 14-week course aimed at educating the top 2–3% of Colonels, Captains (RN), and Group Captains in higher command and operational art."[74] The course, also available for flag officers, is, "in the light of recent developments [in operations in Afghanistan and Iraq], . . . primarily focused on the military-strategic and operational levels set in the wider strategic context."[75]

Adaptability Training

The British Army does not train units to adapt; rather, it *adapts the training.* Instead of trying to prepare units to be ready for anything, OPTAG and similar organizations train units for the specific operational environment they will face. Like the U.S. Army, the British Army adapts predeployment training as the situation evolves. Even though the British Army predeployment-training system does not aim to produce highly adaptable units, it does allow the British Army as a whole to adapt to a wide range of operational environments. The force-preparation regime ensures that British units are highly trained in the skills they will need in any and all contingencies and provides a foundation on which to adapt units to a specific operational envi-

---

[74] Defence Academy of the United Kingdom, "Higher Command and Staff Course," Web page, 2009.

[75] Defence Academy of the United Kingdom, "Higher Command and Staff Course."

ronment. Predeployment training then puts the finishing gloss on this adaptive foundation. Indeed, as previously noted, British doctrine recognizes an explicit requirement for tailored training. Thus, the British Army model falls squarely within the parameters of ongoing adaptation based on lessons learned.

The Train, Advise, and Assist Mission in the UK

In addition to investigating predeployment training for operational missions, we were asked to consider the manner in which British forces are trained for advisory and training missions with third countries. The UK is an interesting case in this regard because it has a relatively robust TAA footprint and focuses considerable attention on building the capacity of less-capable partner countries around the world. This section considers the following aspects of the UK's TAA mission:

- the selection of advisers and trainers
- how organizations conduct the training
- which specific skills are trained
- where the trainers are deployed
- how training is assessed and the nature of the lessons-learned process
- key distinguishing features.

The Selection of Advisers and Trainers

In the UK, there are two tracks—advisers and training teams—for conducting TAA missions abroad. It is not entirely clear whether these TAA missions are career-enhancing or not. What is clear, however, is that there appears to be no expectation that individuals will, at some stage in their military careers, deploy for a TAA mission. (The opposite is true in, for example, France.) With a shortage of forces to take up new rotations to Iraq and Afghanistan, it appears that the UK's approach to selecting trainers and advisers has been mostly ad hoc or based on availability rather than trying to pinpoint individuals with advisory experience or time in country as trainers.

How Organizations Conduct the Training and Skills Taught

The British TAA concept has evolved over the following missions over roughly the past 30 years:

- the employment of British Advisory Training Teams in Oman in the 1970s
- the development of British Military Advisory Training Teams in the 1980s
- the employment of the International Military Advisory Training Team Sierra Leone in 2001
- Operation Mentoring Liaisons in Afghanistan in 2005
- Operation Monogram counterterrorism capacity-building in 2005.

The most significant change has occurred in the type and location of training offered. In the 1970s, 1980s, and 1990s, for example, training emphasized defense reform and military professionalization in regions where the UK had direct colonial ties. Since 9/11, TAA missions have emphasized (1) training larger numbers of forces as part of international teams and (2) conducting bilateral training to build partner capacity for counterterrorism in regions where the UK does not necessarily have past colonial connections.

The UK's main venue for predeployment training of foreign trainers and advisers is OPTAG, which is focused on operational predeployment training. OPTAG's mission is to provide appropriate specialized training to forces deploying to specific theaters; its foundation is predeployment training for Northern Ireland. OPTAG provides both individual and unit-level training.[76]

A two-week predeployment training course is the norm for senior officers (for units, the training lasts much longer). The two-week course curriculum includes courses on

- ROE
- cultural awareness
- basic language skills

---

[76] Each service in the UK has its own version of OPTAG training.

- mine awareness
- helicopter awareness and safety
- basic media awareness.

All British trainers headed for Iraq or Afghanistan receive OPTAG training, and that training remains current for only six months. To date, no British civilians have attended OPTAG, although they may in the future. The only non-British forces to have attended OPTAG have come from Canada and Australia.

Most of OPTAG's current cadre of 190 trainers are posted to locations in the UK to conduct training. A smaller number are deployed in theater at any given time. OPTAG offers individual reinforcement courses for soldiers who are not part of a unit package. All courses culminate in a realistic simulation exercise using members of relevant local diaspora communities to add realism.

Where Trainers and Advisers Are Deployed

At present, OPTAG training is primarily focused on predeployment training for forces headed to Afghanistan. In addition to OPTAG training, however the MOD conducts several training exercises in support of British missions in Africa, the Caribbean, and Eastern Europe. Moreover, there are a British security and advisory team in the eastern Caribbean, which provides counternarcotics-related advice, training, and assistance; a British peace-support team in South Africa, which provides a combined peace-support–operations capability; and Exercise Green Eagle in Sierra Leone.

How Training Is Assessed and the Nature of the Lessons-Learned Process

The British use the term *lessons identified* rather than *lessons learned* because they believe that until lessons are internalized and put into practice, they are only identified. In general, the process for collecting lessons in the UK is fairly robust. However, processes for analyzing, validating, and disseminating lessons for TAA missions in particular could be improved. Collecting lessons identified is a high priority for OPTAG, which often collects specific lessons during the frequent visits

its trainers make to the field to assess the effectiveness of OPTAG's predeployment training. The insights gained from these visits feed into the training process at the beginning of the OPTAG cycle shown in Figure 4.7. Lessons can be identified at many steps along the way: during training itself (i.e., during cascade training and confirmatory training[77]), during the mission-rehearsal exercise, or via the training reports that are provided by the individual units.

Key Distinguishing Features of the British TAA Approach

There are several distinguishing features of the UK's approach to building partner capacity through TAA missions. First, there is no clear end state identified for partner-capacity-building activities, and there is no grand strategic plan or vision for partner countries. Second, there is no human-rights vetting process at the individual level; such a process exists only at the country and organizational levels.[78] Third, priority countries for capacity-building counterterrorism training—and, indeed, for all British training with third countries—are determined by the Contact Group, which includes officials from the MOD, the FCO, and the Department for International Development.

The UK is trying to change the "fair-weather-friend" reputation it has earned in its partner countries by conducting recurring training, creating memoranda of understanding, and carrying out needs assessments. However, it is not forcing those countries to adopt British approaches and mindsets. In short, the British approach is about providing partners with the tools they need to address their own security challenges, not about forcing British methods on them.

---

[77] Cascade training occurs over a period of 10 full working days and is arranged by the units. Confirmatory training lasts for seven days and is a battle-group exercise focused at the subunit or subplatoon level. It emphasizes reaction to incidents in a tactical environment and includes aviation, air, or indirect fire (discussions with OPTAG officials, Folkestone, Kent, February 2008).

[78] The Foreign and Commonwealth Office conducts human rights vetting for the UK government.

Key Insights

There are several insights from British practice that deserve attention from the United States:

- The UK believes that short-notice joint operations require more than simply establishing a joint staff in a joint HQ. To be fully effective and to eliminate duplication of effort, the joint nature of operations is inculcated throughout the individual service structures. Furthermore, the British system shows that a highly joint approach to operations does not require the loss of single-service ethos or identity.
- The UK's use of a readiness profile as the demand function for the single services to provide joint forces seems to work effectively.
- The RN and RAF training models are very similar to U.S. Navy and U.S. Air Force training models.
- The British Army's training methods, which are different from those used by the U.S. Army, are based on the recognition of the difference between force preparation and force generation, and they are organized accordingly. Thus, the British Army has been able to prepare forces for service in a range of operational environments.

The British Army's experience demonstrates that it is possible to prepare effectively for specific operational environments with relatively modest resources. OPTAG requires fewer than 200 service personnel to operate its program and provide a wide range of training. Most of OPTAG's efforts are devoted to training the trainer and involve neither extensive simulation nor expensive instrumentation. This allows the British Army to exploit the capabilities at their CTCs to their fullest. Again, the point of force-preparation training is to ensure that all of a unit's systems are thoroughly exercised under demanding conditions, something that BATUS, BGTU, and the training estate at Sennelager appear to do quite well.

CHAPTER FIVE

India

Introduction

The key to understanding the Indian armed forces' approach to training is their emphasis on schools. Indeed, *training* is virtually synonymous with *schooling* in the lexicon of the Indian military. The training and recruitment chapter in the Indian Ministry of Defence's 2006–2007 *Annual Report* is concerned almost exclusively with the armed forces' various academies, schools, and colleges. There is a course for virtually every significant position an officer might occupy. Preoperational training, which is a fairly high priority for the Indian armed forces, usually consists of a unit cadre attending a school of one sort or another. During Indian Navy work-ups in preparation for operations, training teams administer written quizzes. This emphasis on formally constituted schools is unique among the countries we researched.

In India, collective training is similar to that provided in the United States, but it does not involve CTCs. Indian commanders are responsible for preparing their staffs and subordinate units for wartime missions, which are specific to a geographic assignment. Units train collectively by designing and executing exercises, including field training and command-post training, using largely their own resources. Joint training, however, is episodic, and suffers from an absence of joint structures and joint doctrine. Each of the services has a different approach to warfighting and simply assumes that sister-service approaches are complementary.

The Indian approach appears to have been adequate at the tactical level in recent clashes. Indian forces have been fairly successful in conducting several long-running counterinsurgency (COIN) campaigns not just in Kashmir but also in several other areas of operation, including combat operations during the Kargil War of 1999.

The sources we consulted for this case study were mostly open-source materials. It is very difficult to get Indian officials to discuss the issues this monograph addresses. We were, however, able to consult with a former officer of the Indian Army.

Strategic Demands and Focus

Observers complain that India lacks a national-security strategy. Retired Brigadier Arun Sahgal, a frequent commentator on Indian strategic affairs, has noted that the Indians "have over the years failed to evolve a coherent national security strategy to meet national aspirations."[1] Sahgal's complaint is echoed by outside observers. George Tanham, formerly of RAND, noted that India is a great state characterized by the absence of a national strategy.[2] For all that, India's strategic requirements are derived from the country's strategic environment and aspirations. They are probably no less coherent in the absence of a declaratory strategy than those of the United States, which does have such a strategy.

Strategic Imperatives and Priorities

India faces a diverse array of security challenges both on and within its borders. Pakistan, with whom India has fought three wars, poses a complex hybrid challenge and certainly preoccupies Indian military officials. The provinces of Jammu and Kashmir are the principal areas of contention, and Indian officials believe that Pakistan supports the ongoing insurgency in both. Indian officials also believe that Pakistan,

---

[1]   Arun Sahgal, "National Military Aspirations and Military Capabilities: An Approach," in Vijay Oberoi, ed., *Army 2020: Shape, Size, Structure and General Doctrine for Emerging Challenges*, New Delhi: Knowledge World, 2005, p. 103.

[2]   Stephen Philip Cohen, *India: Emerging Power*, Washington, D.C.: Brookings Institution Press, 2001, p. 63.

either actively or tacitly, supports such acts of terrorism as the December 2002 attack on the Indian parliament. Although Pakistan's support of irregular warfare poses the most likely challenge, Pakistan's large and well-equipped conventional forces cannot be ignored. To the northeast, Indian officials view China's rapid military modernization with some trepidation, although relations between the two countries are relatively close and cordial. India must also remain concerned about instability in such neighboring countries as Nepal, Bhutan, Bangladesh, Afghanistan, and Sri Lanka. In the maritime domain, India is situated in the middle of the world's busiest and most constricted trade routes, including the Straits of Hormuz to the west and the Straits of Malacca to the east. Internally, India must deal with a number of insurgencies proceeding at varying levels of intensity. In addition to ongoing problems in Jammu and Kashmir, insurgency troubles the states of Assam, Nagaland, Manipur, and Tripura, not to mention several other states that are experiencing lower levels of violence and terrorism. Moreover, Indian security forces must cope with countrywide terrorism and low-level violence under the aegis of the Maoist-oriented Naxal Movement. Finally, the Indian subcontinent is perennially the site of some of the worst natural disasters in the world.[3]

India's strategic imperatives extend, however, beyond meeting these challenges. According to Stephen Philip Cohen, there is an idealistic tone to India's foreign policy that stems from Gandhian idealism. This idealism is one of the reasons that India is one of the largest supporters of UN peacekeeping operations. Indian strategists are also keenly aware that their international status, especially within the developing world, depends on their idealistic behavior. Successive Indian governments have aspired to great-power status in one form or another, an ambition based on Indians' sense of their history, on their ideology, and on their emerging economic power. According to Cohen, Indians consider their country a status quo power entitled to regional hegemony. This sense of entitlement is tempered somewhat

---

[3] Ministry of Defence, Government of India, *Annual Report, 2006–2007*, undated, pp. 2–8, describes the general security situation; a discussion of internal insurgency challenges appears on pp. 20–23.

by their awareness of their limited ability to enforce their hegemonic aspirations.[4] Therefore, India's strategic imperatives focus on deterring external aggression, mostly from Pakistan; achieving internal stability and security; and enhancing India's international status.

The Role of Operational Deployments in National Strategy

Because India's security challenges lie either on or within its borders, power projection does not play a very important role in Indian defense strategy. For the most part, out-of-country deployments take place under UN auspices in support of UN peacekeeping operations. Therefore, such deployments do not directly affect Indian security. India's one major unilateral deployment, to Sri Lanka between 1987 and 1990, went awry, leaving Indian officials with a deep reluctance to intervene in another country's affairs.[5]

Indian forces do, however, regularly deploy within India from their peacetime garrison locations to either serve in ongoing COIN operations or increase readiness along India's borders. The Indian Air Force and, especially, the Indian Navy also deploy to conduct bilateral exercises whose apparent goal is to strengthen India's relations with potential partners.[6]

How India's Strategic Imperatives Compare with Those of the United States

At first glance, India's strategic context and imperatives contrast sharply with those of the United States. India faces significant conventional challenges on its borders, whereas the United States' borders with Canada and Mexico are not militarized. India faces considerable internal instability, whereas the United States has not experienced violent internal conflict for over a century. The United States is the sole remaining superpower, whereas India struggles to attain regional hegemony. Consequently, although almost all of the United States' scenar-

---

[4]   Cohen, *India,* pp. 64–65.

[5]   Ministry of Defence, Government of India, *Annual Report, 2006–2007,* pp. 23–24.

[6]   Ministry of Defence, Government of India, *Annual Report, 2006–2007,* pp. 29–32, 38–39.

ios for the employment of its military forces involve power projection, almost none of India's do.

At a higher level of abstraction, however, India's strategic imperatives resemble those of the United States. Like the United States, India must prepare its forces to conduct ongoing COIN operations even as it hedges against the outbreak of MCO. Like the United States, the Indian armed forces are organized, trained, and equipped for combat operations but must routinely and continuously adapt themselves to conduct COIN successfully.

The Defense Establishment

The Components

The Indian Army. With approximately 1.1 million members, the Indian Army is by far the largest of India's military services and receives the lion's share (approximately 48 percent) of the defense budget.

The Indian Navy. With 55,000 personnel and a little over 18 percent of the defense budget, the Indian Navy is the smallest of the three services. The Indian Navy aspires to become a power-projection force but is limited for the moment by its aging inventory of ships.[7]

The Indian Air Force. The Indian Air Force has approximately 170,000 airmen and is the fifth-largest air force in the world, according to *Jane's International Defence Review*. It is quite capable. In Cope India '04, a joint Indo-U.S. air exercise, the Indian Air Force effectively fought its U.S. competitors to a draw.[8]

Human Resources

The separate Indian services promote officers and enlisted soldiers according to their own independent practices. In general, the Indian Army and the Indian Navy tend to emphasize seniority, and the Indian Air Force provides more opportunities for rapid promotion.

---

[7] "Navy, India," *Jane's Sentinel Security Assessment—South Asia*, January 3, 2008; Ministry of Defence, Government of India, *Annual Report, 2006–2007*, pp. 15, 28–34.

[8] "Air Force, India," *Jane's Sentinel Security Assessment—South Asia*, February 13, 2008.

Furthermore, as is the case in many of the armies we researched, Indian Army officers tend to be senior to their U.S. counterparts both in age and in terms of the rank at which they command.[9]

The Indian armed forces are composed of long-serving volunteers. Enlisted members still sign on for an initial term of 17 years in spite of various proposals to reduce the term of service to seven years as a means of increasing recruitment and decreasing pension expenses. A *naik*, or corporal, remains in service for up to 24 years or until age 49, whichever comes first; a *subedar major*, or sergeant major, serves for up to 36 years or until age 54.[10]

Professional military education in the Indian armed forces is comparable to that provided in the U.S. armed forces. Indian officers attend some sort of school to prepare for each distinct phase of their career. Prior to commissioning, Indian officer candidates destined for any one of the three services generally attend the National Defence Academy or the Indian Military Academy. After commissioning, officers attend the branch schools of their specific arms or services. There are also courses to prepare officers selected for command at either the battalion or division levels or at higher echelons. There are also separate courses to prepare officers for staff roles. During approximately their tenth year of service, Indian officers become eligible to attend the Defence Services Staff College, whose purpose is to prepare officers for staff roles at the ranks of major and lieutenant colonel (and their naval and air-force equivalents).[11] Civilian and military leaders identified as potential strategic leaders attend the National Defence College, generally at the rank of colonel or brigadier general. There are also several courses in defense management.[12]

---

[9] Thirteenth Lok Sabha, Standing Committee on Defence, *Manpower Planning and Management Policy in Defence*, August 24, 2001, p. 6.

[10] Thirteenth Lok Sabha, Standing Committee on Defence, *Manpower Planning and Management Policy in Defence*, p. 14.

[11] Inter-Services Institutions, "Defence Services Staff College (DSSC)," Web page, undated.

[12] Ministry of Defence, Government of India, *Annual Report, 2006–2007*, pp. 100–116; National Defence College, home page, undated.

Recruiting and Retention Considerations

As India's economy has grown, concomitantly expanding opportunities for the educated middle class, the attractiveness of military service has declined significantly for many of the same reasons seen in other countries. Families are less willing to accept the frequent dislocations and inconveniences of military life, especially as more-lucrative opportunities beckon in the burgeoning private sector. In 2001, the Indian Army was short 11,000 officers. Although the services have been able to recruit sufficient numbers of enlisted soldiers, the quality of recruits does not appear to have kept pace with requirements.[13]

Manning Strategies

The regimental system is extremely strong in the Indian Army. Most enlisted soldiers, especially in the combat arms, serve their entire career in the same regiment. Much like the British Army, the Indian Army frequently seconds its officers to other duties—including staff duty with brigade and higher-level HQ and instructor duty at branch and specialty schools—as they increase in rank. Nonetheless, except when so detailed, officers remain with their regiments. Regiments regularly rotate, as units, between peacetime postings and more-active service either in support of COIN operations or in garrisons along India's borders.

At various times throughout their careers, soldiers are liable to seconding to one of the special units, such as the Rashtriya Rifles or the Assam Rifles, that are permanently engaged in COIN operations. The Indian Army mans these units through an individual-replacement system, and the express goal is to maintain continuity in these extremely complex and highly sensitive operations. Officers and men from all branches, not just the infantry, are detailed to serve in these units.[14]

---

[13] "India Tackles Rising Army Personnel Crisis," *Jane's Defence Weekly*, January 3, 2007. See also Rahul K. Bhohsle, "India's National Aspirations and Military Capabilities: A Prognostic Survey," in Oberoi, *Army 2020*, p. 152. See Fourteenth Lok Sabha, Standing Committee on Defence, *Demands for Grants (2007–2008)*, April 2007, p. 118, for analysis of Air Force attrition.

[14] Discussions with a former Indian Army officer, Arlington, Va., November 6, 2007.

Priorities: The Mission Set and the Range-of-Operations Focus

With the exception of the Indian Army, the Indian defense establishment lacks much in the way of formal doctrine. Nonetheless, it is possible to discern the Indian armed forces' range of operations from their actual practice. In general, that practice spans the full range of operations. The Indian Army recognizes a range of military operations from humanitarian assistance to nuclear war. The Indian Air Force and the Indian Navy focus mostly on the high end of the spectrum, particularly conventional war with Pakistan, although they have supported operations at the low end (e.g., they have supported civil authorities in the wake of a national disaster).[15]

An Assessment of Component Roles Against Mission Sets

The Indian Army

With the exception of nuclear deterrence, the Indian Army is directly responsible for most of India's pressing strategic priorities. In its primary role, the Indian Army must prepare to undertake ground operations to retain Indian territory or to compel an enemy to accept Indian terms. In its secondary role, the Indian Army supports civil authorities in confronting insurgents and terrorists, but, in practice, it must frequently take the lead in this role. Additionally, the Indian Army also prepares to provide assistance to civil authorities in case of emergency. Finally, the Indian Army provides most of the forces for peacekeeping operations.[16]

The Indian Navy

The Indian Navy is mostly responsible for maritime security in and around the subcontinent and for conducting military diplomacy by means of joint and combined exercises with foreign navies. Eventually,

---

[15] Indian Army, Army Training Command, *Indian Army Doctrine*, October 2004, p. 12. For descriptions of Indian Navy relief operations, see Ministry of Defence, Government of India, *Annual Report, 2006–2007*, p. 32.

[16] Indian Army, Army Training Command, *Indian Army Doctrine*, p. 9.

it will deploy missile-firing submarines as part of the country's nuclear deterrent.

The Indian Air Force

The Indian Air Force has three principal roles: air support of ground forces, operational interdiction, and nuclear deterrence. Even though these roles involve the support of ground operations, the Indian Air Force has resisted both greater integration with the Indian Army and subordination to a joint staff.[17]

Specialty Forces

The Indian government maintains a number of paramilitary formations for combating civil disorder and insurgency. We have already mentioned the Rashtriya Rifles, which are under the Indian Army's jurisdiction. The Rashtriya Rifles alone accounted for 63 battalions of the Indian Army in 2007. There are several similar organizations, including the Assam Rifles, the Border Security Force, the Central Industrial Security Force, the Central Reserve Police Force, the Rapid Action Force, and the Special Protection Group. According to Indian government policy, these paramilitary forces have primary responsibility for COIN, but, in practice, they are not always adequate to the task.[18]

Training Regimes

Overall Training Methodology

Like many of the other countries we researched, India does not have a single training methodology. Joint structures are weak, joint doctrine is

---

[17] The Indian Army has been quite sensitive about this issue. See, for example, Thakur Kuldip S. Ludra, *Air Land Battle—The Indian Dichotomy: A Report Submitted to the Joint Chiefs of Staff Committee*, New Delhi: Institute for Strategic Research and Analysis, 2001, p. 4; V. K. Shrivastava, "Indian Air Force in the Years Ahead: An Army View," *Strategic Analysis*, Vol. 25, No. 8, November 2001, pp. 937–944.

[18] Omar Khalidi, *Khaki and the Ethnic Violence in India: Army, Police and Paramilitary Forces During Communal Riots*, Gurgaon, India: Three Essays Collective, 2003, Table 1; Ministry of Defence, Government of India, *Annual Report, 2006–2007*, p. 23.

nonexistent, and joint training is episodic, though robust when it does occur. To describe Indian training methodologies, we must therefore describe the methodologies of the three single services. None of the services has established a CTC. Instead, collective training largely takes the form of service or joint exercises.

Overall, however, the Indian armed services share one common trait: the tendency to equate training with schooling. The recruitment and training chapter of the Ministry of Defence's 2006–2007 *Annual Report* describes the Indian armed forces' various schools and centers, not collective training. In India, every important Indian Army center is called a school. Courses at the various schools, including the High Altitude Warfare School, the Counter Insurgency and Jungle Warfare School, and the various corps battle schools, include a substantial component of classroom instruction, although this is augmented with a healthy dose of practical instruction.

Evaluated in the light of Indian operations during the 1999 Kargil War and the 2002 general mobilization during Operation Parakam, these training methodologies produced results much like one might expect. At the tactical level, the Indian armed forces suffered perhaps unnecessary casualties during soldiers' first exposure to combat. At the operational level, there was considerable difficulty integrating air operations and ground maneuver. At the higher levels of command, there was considerable difficulty controlling the maneuver of divisions and corps. That said, the Indians have since worked through these problems. It should be noted, however, that, although there is some inefficiency at every level, the Indian armed forces have consistently demonstrated a high level of effectiveness and, at the level of the force, a high level of adaptability.

The Indian Army. As previously noted, the Indian Army must train forces both for MCO against one of the country's neighbors and for the conduct of ongoing operations. In general, training for the former is left to operational HQ, which use their own resources to train subordinate echelons and units and employ traditional approaches, such as maneuvers and CPXs. In a manner not unlike the U.S. system described in CJCSM 3500.03A, *Joint Training Manual for the Armed Forces of the United States*, Indian commanders focus their units on the tasks they anticipate having to perform and the conditions they think they will face in combat. At least one observer has noted that these efforts lack realism and utility below the battalion level. According to Lieutenant General N. S. Naskari, India has

> been paying lip-service to sub-unit training at best[,] retaining 'cadre' type . . . training, . . . despite the fact that from Leh to Lungleh, our army has fought 90 percent of battles/conflicts, in both wars and . . . [in COIN] situations, at sub-unit and unit levels. If we have had success, then it is more due to the commitment and courage of our junior leaders than the systematic training at the unit level. . . . Our field firing exercises and battle inoculations are a joke.[19]

Although the Indian Army has announced (probably in response to such sentiments) its intention to improve its training infrastructure, it has yet to establish anything like the U.S. CTCs.

Preoperational Training

The Indian Army. The Indian Army recognizes an imperative to train contingency forces for their operational environment, and it has established an extensive infrastructure for that purpose.[20] The Indian Army's various centers and schools form the heart of the service's approach to preparing forces for ongoing operations. These schools include the Counter Insurgency and Jungle Warfare School, the Centre for United Nations Peacekeeping Operations (a collective project of the Indian Army, the Ministry of External Affairs, and the quasi-official United Services Institute), the High Altitude Warfare School at Gulmarg, and various corps battle schools. These schools focus on preparing Indian Army units for the specific operations they will undertake. Instructors at the corps battle schools in particular are seconded from units that have just finished an operational tour in the relevant area of

---

[19] N. S. Naskari, "Doctrinal Changes and Imperatives of Force Restructuring," in Oberoi, *Army 2020*, p. 232.

[20] For the Indian Army's doctrinal imperative, see Indian Army, Army Training Command, *Indian Army Doctrine*, Part 3, p. 13.

operations. This practice ensures that information on the enemy, the population, and the conditions of operations is current.[21]

According to the Indian Army's *Doctrine for Sub-Conventional Operations*, preparation for an operational deployment proceeds through three phases. In the first, the orientation phase, the unit sends its training team to the relevant schools for an intense period of training designed to enable the training-team members to train their units. These personnel then train their units under cadre supervision at the relevant corps battle school for six weeks. During this period, the unit actually conducts patrols under the supervision of school staff. Finally, the unit undergoes "on-the-job training" for three to four weeks as it gradually takes over responsibility from its predecessor.[22]

The Indian Navy. The Indian Navy's preparation for operations closely resembles that conducted by the RN and takes place under the direction of the Indian Navy's FOST, an organization established in 1992. FOST comprehends three teams. There are two teams for training the crews of ships that are at or below the corvette class, and there is one team for training all the larger ships in the Indian Navy. Operational sea training proceeds through four phases that begin in port and extend through single-ship, multiship, and carrier–battle-group operations. Written examinations are an important part of the work-up process.[23]

Joint Structures and Training for Joint Operations

Joint structures in the Indian armed forces are best described as immature. The Indian government did not establish its Integrated Defence Staff, analogous to the U.S. Joint Staff, until 2001. Originally, the Integrated Defence Staff was to be headed by a chief of defense, a uniformed officer to whom the three service chiefs would be at least nominally subordinate. Currently, however, the organization is headed by a lieutenant general and functions as an advisory committee to the collective chiefs of staff. Like the U.S. Joint Staff, it lacks any authority to

---

[21] Discussions with former Indian Army officers, Arlington, Va., January 25, 2008.

[22] Indian Army, *Doctrine for Sub Conventional Operations*, December 2006, p. 51.

[23] Indian Navy, "Flag Officer Sea Training," Web site, undated.

direct the services.[24] Moreover, because of its relative novelty, it lacks the considerable de facto influence possessed by its U.S. counterpart.

For this and other reasons, joint planning, force development, and training remain quite rudimentary. Sahgal noted that "even after four wars and innumerable crises . . . [India has] failed to evolve joint doctrine and concepts"; instead, each service has tended to evolve separate warfighting concepts that assume support from the other services.[25]

Joint training does take place, however, although it mostly occurs at the instigation of one of the services (usually the Indian Army). This joint training mainly takes the form of major exercises. To test the Indian Army's "Cold Start" doctrine, the Indian armed forces have conducted five joint exercises of varying sizes since 2004. In his discussion of the evolution of the Cold Start doctrine, however, Walter Ladwig notes that "despite multiple rehearsals, the two services [the Indian Army and the Indian Air Force] consistently failed to integrate their actions" in these war games.[26] According to Ladwig, this disjuncture resulted in no small part from the fact that the Indian Air Force, which disagreed with the underlying concept, preferred to wage war independently in the aerospace domain. Absent more-robust joint structures, Ladwig does not foresee Indian improvements in joint operations and training.[27]

Assessing Indian Training Effectiveness: Kargil 1999

It is somewhat difficult to assess the effectiveness of Indian training methods after the 1999 Kargil War, the most recent significant operation involving Indian forces. Since that war, the Indian armed forces have initiated several reforms that ought to have improved their effectiveness. India's performance in Kargil demonstrated that the Indian armed forces were capable of adapting to the unique demands of this

---

[24] Integrated Defence Staff–India, "Welcome Message from Chief of Integrated Defence Staff," Web page, undated.

[25] Sahgal, "National Military Aspirations and Military Capabilities," p. 105.

[26] Walter C. Ladwig III, "A Hot Start for Cold Wars: The Indian Army's New Limited War Doctrine," *International Security*, Vol. 32, No. 3, Winter 2007–2008, pp. 158–190.

[27] Ladwig, "A Hot Start for Cold Wars," pp. 158–190.

operational environment, but it also revealed that individual units did not display the same degree of adaptability.

In the fall of 1998, Pakistani officials decided to launch a limited offensive into Indian Kashmir in the coming winter. Before then, Indian forces had generally withdrawn from the inhospitable terrain, which ranges in altitude from 16,000 ft to 18,000 ft. During the period of Indian inactivity, the Pakistanis infiltrated soldiers of their Northern Light Infantry (in mufti), complemented by supporting arms and so-called mujaheddin, into commanding positions in the mountains, where they subsequently entrenched. Eventually, about 1,700 Pakistani combatants occupied a front of approximately 150 km. Apparently, the Pakistanis bet that the relative impregnability of their position, coupled with their country's nuclear capability, would compel India to accept this entrenchment as a fait accompli.[28]

The Indian XVth Corps discovered the intrusion in early May 1999, and Indian Army forces who were on hand because they were engaged in COIN duty immediately set out to recapture the position. These initial attacks failed. The Indian troops, who had been fighting in tropical heat, were not prepared for arctic fighting at high altitude, and they suffered tremendously from altitude sickness and exposure. Thinking they faced only lightly armed guerillas, the local Indian commanders attempted frontal attacks while they were virtually unsupported by either artillery or air power. When the Indian Air Force did become involved, it proved similarly unprepared for operations at high altitude. A major Indian attack on the Tololing Hill complex on May 22, 1999, failed.

Indian forces adapted. The Indian Air Force decided to shift tactics by attacking Pakistani supply lines. Because it was difficult to strike specific targets, the Indian Air Force instead used air strikes to cause avalanches that rendered trails impossible and buried supply caches. Using laser-guided munitions, Mirage 2000 aircraft had some success

---

[28] Unless otherwise noted, our description of the events of the Kargil crisis are from Marcus P. Acosta's excellent article, "The Kargil Conflict: Waging War in the Himalayas," *Small Wars & Insurgencies*, Vol. 18, No. 3, September 2007, pp. 397–415. Sharif's and Musharraf's involvement is confirmed in Shaukat Qadir, "An Analysis of the Kargil Conflict 1999," *RUSI Journal*, April 2002.

against specific positions, but the most-effective fire support came from artillery. The Indian Army vastly increased the amount of artillery supporting this campaign, deploying as many as 20 batteries and adapting tactics to the particular nature of high-altitude combat. Perhaps most importantly, the Indian Army deployed the 2nd Rajputana Rifles under the command of Lieutenant Colonel M. B. Ravindranath. Colonel Ravindranath, a graduate of the High Altitude Warfare School, prepared his unit thoroughly for mountain combat at night, requisitioning cold-weather clothing, acclimatizing his men rapidly and in stages, and establishing the necessary stockpiles and logistics for an assault. Augmented by staff from the High Altitude Warfare School and supported by over 120 guns of various types, the 2nd Rajputana Rifles seized Tololing Hill on June 20, 1999. By July 26, 1999, Indian forces had cleared the mountains of Pakistani invaders.

Although the Indians adapted successfully as a force, individual units did not. Infantry units that had prepared for COIN operations proved woefully unprepared for mountain combat. Commanders failed to employ supporting arms effectively. From the air perspective, apparently, no one had thought through, let alone practiced, supporting ground operations in high-altitude terrain.

Understanding how the Indians did adapt is key. First, they found a commander who had been to the right school, and they supplied him with personnel with the right skills. They changed the composition of the force and adapted their tactics, especially their infantry tactics. They also adapted their fire-support techniques to make them more effective in the mountainous terrain. In short, the Indians prevailed at the tactical level because they had an adaptable system, not because they had adaptable units.

Although the Indians were able to adapt successfully to the problem at hand, they were not able to solve deeper, systemic issues during the campaign. Joint C2, in particular, remained a problem. According to an anonymous officer on the staff of one of the divisions involved, units only learned about air activity when the aircraft were overhead. They were seldom informed of the air strikes' intended targets and results.[29]

---

[29] Ludra, *Air Land Battle–The Indian Dichotomy,* p. 6.

Training for Coalition Operations

For the most part, Indian training with other countries takes the form of military diplomacy and is conducted mostly by the Indian Navy and the Indian Air Force. The Indian Navy and the Indian Air Force exercise extensively with countries in the region and with such Western powers as the United States, France, and the UK. Indeed, it is not an exaggeration to say that the Indian Navy and the Indian Air Force exercise more extensively with foreign militaries than with the Indian Army.[30] It is unlikely, however, that any foreign powers will intervene directly in any future Indian war, and Indian Army training reflects this fact.

Training Methodologies for Foreign Militaries

In general, India's methods for training foreign militaries are similar to those of the United States. The Indian armed forces train foreign militaries by including foreign students in their various schools, much as occurs in the U.S. International Military Education and Training program. Providing English instruction to these foreign students is often necessary to enable them to participate in Indian courses. India's Centre for UN Peacekeeping explicitly orients its curriculum on foreign officers and aims to be an international center of excellence in this field. Indeed, U.S. service members have attended both the High Altitude Warfare School and the Counter Insurgency and Jungle Warfare School.[31]

The Indian armed forces also maintain training teams in a number of countries, especially in Africa. These missions mirror the Indian armed forces' emphasis on formal schools. For example, the centerpiece of the Indian military training team in Bhutan is the Wangchuk Lo Dzong Military School, which essentially prepares Bhutanese soldiers to attend Indian military schools. Similarly, the Indian mission to Laos consists of four soldiers assigned as instructors at the Laotian military academy.[32]

---

[30] Ministry of Defence, Government of India, *Annual Report, 2006–2007*, pp. 33, 38.

[31] Ken Denny, "Alaskans Train at Top Jungle Warfare School in India," *National Guard Bureau News*, June 21, 2004.

[32] Ministry of Defence, Government of India, *Annual Report, 2006–2007*, pp. 152–156. Our description of the Indian training team's missions in Bhutan is from Indian Military Training Team in Bhutan, home page, undated. Our description of the Indian training

Comparison with U.S. Regimes

The Indian armed forces' maneuver-training regime is similar to that of the United States, but it is not nearly as advanced. To prepare for combat operations, the Indian armed forces follow practices similar to those described in CJCSM 3500.03A, *Joint Training Manual for the Armed Forces of the United States*: identifying likely wartime tasks and conditions and then developing and conducting training events to practice those tasks. True joint-maneuver training is still episodic and would in any case be difficult in the absence of any standing joint operational HQ.

The Indian Army's reliance on schools to prepare forces for operations is the principal difference between the U.S. and Indian regimes. Much like OPTAG in the British system, these schools constitute tangible acknowledgment by the Indians that the conditions of irregular warfare require painstaking collective preparation for the operational environment, particularly at the battalion level and below. Second, in contrast to the U.S. system of preparing units for operations through collective maneuver training, the Indian reliance on schools emphasizes preparing units by preparing leaders. In this model, the leader's understanding of the operational environment and its dynamics is far more important than unit cohesion. Indeed, the importance the Indian armed forces attach to leader capability is demonstrated in their use of an individual-replacement policy in their premier COIN force, the Rashtriya Rifles.

Adaptability Training

From the foregoing discussion, it should be clear that the Indian armed forces do not train individuals and units to adapt. Rather, the Indian armed forces adapt the training. By using officers and NCOs who have just completed a tour in the area of operations as cadres at the Counter Insurgency and Jungle Warfare School and at the various corps battle schools, the Indian Army continuously adapts unit training to meet the demands of the operational environment.

---

team's mission in Laos is from Embassy of India in Lao PDR, "Assistance to Lao PDR," Web page, 2007.

Key Insights

Like the United States, France, and the UK, India has found it necessary to tailor training for irregular warfare to specific operational environments. This is even more significant in India's case because the country must conduct COIN in a variety of different operational environments (ranging from the mountainous terrain in Jammu and Kashmir to the jungles of Nagaland) and in India's varied cultural and physical geography. That India, like these other countries, has not attempted to train individuals and soldiers for adaptability does not necessarily prove that training for adaptability is either impossible or undesirable. However, it does indicate that systemic adaptation is a reliable means of preparing forces for irregular warfare and other complex environments. Given India's long, varied, and relatively successful history of conducting COIN in various regions, this finding should carry special weight.

Second, leader training seems to be an especially cost-effective way of preparing forces for operations. India's principal means of preparing forces for COIN operations involves schools that seem to closely resemble OPTAG, although they are more intense. The schools train unit-training teams, which then train their units under the close supervision of school instructors. By all accounts, such training has resulted in a high degree of situational understanding, an extremely low incidence of abuse and collateral damage, and corresponding increases in security. In more-intense operations, such as the 1999 Kargil War, Indian leaders have been able to adapt tactics and organizations to the circumstances and successfully conduct operations for which they had not recently prepared.

CHAPTER SIX

Israel

Introduction

Israel faces more immediate and serious threats to its security than the other states considered in this monograph. It is small, both in terms of land mass and population, relative to its adversaries. It shares borders with openly hostile states and is situated among a number of countries that refuse to acknowledge it as a state. In addition, several nonstate adversaries have the ability to attack Israeli citizens.

Israel also differs from the other countries we assessed in that it has recently faced a direct threat to its security and achieved less-than-ideal results. In summer 2006, Israel confronted Hezbollah in Lebanon. Provoked by the abduction of two members of the Israel Defense Forces (IDF) and a barrage of rockets launched from Lebanon, Israel struck back at Hezbollah. The IDF was able to destroy many long- and intermediate-range rocket launchers, and it inflicted significant damage on Hezbollah. However, after a 34-day ground and air campaign, the IDF was unable to stop the nonstate adversary from launching short-range rockets into Israel. Much of current Israeli security policy can be viewed as a reaction to difficulties encountered during the 2006 Second Lebanon War. The difficulties Israel faced in Lebanon in 2006 and the perception that the IDF failed are a fundamental security issue for Israel because the country's ability to deter aggression is in no small part based on maintaining among Israel's adversaries the notion that the IDF is invincible. That image was tarnished during the Second Lebanon War.

Israeli President Shimon Peres once described the range of threats confronting Israel by saying that the country needed to prepare itself for attacks from "knives, tanks, and missiles."[1] By *knives*, he meant the threat of nonstate adversaries; today, Israel faces such threats from Hezbollah, Fatah, and Hamas. *Tanks* refers to conventional military threats, such as Syria. By *missiles*, President Peres meant the threats associated with Iran and other groups that might turn to weapons of mass destruction.

The fact that Israel must prepare its military for a variety of threats (threats the Israelis call the *rainbow of operations*) makes the country a good point of comparison with the United States. The small size of the IDF and the fiscal limits under which it operates have forced Israeli defense planners to make some difficult choices. The disappointing performance of the IDF in the Second Lebanon War raises questions about the wisdom of these choices.

Some have identified Hezbollah's performance in the 2006 conflict as a harbinger of the type of threat that the United States may face in the future.[2] The Second Lebanon War posed a new kind of challenge for Israel in that the conflict was

> a limited war of a state against a non-state actor operating from the territory of a failed state that does not control its own territory. The nonstate player fought as a guerilla force, though in some areas it possessed state-like capabilities, acquired from supporting states. For example, Hizbollah had various kinds of guided missiles: anti-tank, anti-aircraft, and land-to-sea missiles as well as assault UAVs [unmanned aerial vehicles], and had the ability to strike deep in Israel's home front.[3]

---

[1] This quote was used in the title of a book that discusses the Israeli military and the revolution in military affairs. See Eliot A. Cohen, Michael J. Eisenstadt, and Andrew J. Bacevich, *Knives, Tanks and Missiles: Israel's Security Revolution*, Washington, D.C.: Washington Institute for Near East Policy, 1998.

[2] Frank G. Hoffman, "Hizbollah and Hybrid Wars: U.S. Should Take Hard Lesson From Lebanon," *Defense News*, August 14, 2006, p. 52; Frank G. Hoffman, "Neo-Classical Counterinsurgency?" *Parameters*, Summer 2007, pp. 71–87.

[3] Shlomo Brom, "Political and Military Objectives in a Limited War Against a Guerilla [sic] Organization," in Shlomo Brom and Meir Elran, eds., *The Second Lebanon War: Strategic Perspectives*, Tel Aviv: Institute for National Security Studies, 2007, pp. 13–14.

During the conflict, Hezbollah posed a strategic challenge to Israel by launching rockets into Israeli territory. At the beginning of the war, Hezbollah had a large stockpile of rockets, as shown in Table 6.1.

Although the Israeli Air Force (IAF) was effective against medium- and long-range launchers, "it was hard pressed to attack short range rockets with any measure of success,"[4] and 100–200 Hezbollah Katyusha rockets hit Israeli communities before the ceasefire ended the war.[5] At the operational and tactical levels, Hezbollah also proved to be a difficult challenge for the IDF. Tactically and operationally, Hezbollah proved a tenacious opponent for the Israeli Army, which was largely unprepared for what it found in its initial offensive into southern Lebanon:

> The resistance carried out by the highly professional and well-equipped guerrilla fighters proved a major challenge to IDF units. The IDF had to deal with an intricately camouflaged and reinforced foxhole and tunnel system through which Hizbullah fighters carried out deadly ambush attacks. Hizbullah preparations for war were attested by the fact that it had carved up South Lebanon into over 170 combat quadrants managed from . . . [approximately] 50 scattered command bunkers. This bunker network, situated in what many call the 'Triangle of Death' given its dense

Table 6.1
Hezbollah Rocket Resources at the Beginning of the Second Lebanon War

Type	Range (km)	Payload (kg)	Quantity
122-mm Katyusha	7–40	7	13,000
220-mm and 302-mm Fadjr-5 and Fadjr-3	45–70	50–175	~1,000
Zelzal 2	200	400–600	Dozens

SOURCE: Giora Romm, "A Test of Rival Strategies: Two Ships Passing in the Night," in Brom and Elran, *The Second Lebanon War*, p. 53.

---

[4] Gabriel Siboni, "High Trajectory Weapons and Guerilla [sic] Warfare: Adjusting Fundamental Security Concepts," *Strategic Assessment*, Vol. 10, No. 4, February 2008, p. 17.

[5] Romm, "A Test of Rival Strategies: Two Ships Passing in the Night," p. 57.

vegetation and deep crevices that allow for deadly ambushes, incorporated war rooms equipped with the best hi-tech instruments such as computers, . . . electronics and night-vision equipment. Many IDF units found it difficult to operate amongst this bunker network as they had not received appropriate training for combat against camouflaged bunkers.[6]

Clearly, Hezbollah was a very different challenge for the IDF as a whole and, in particular, for the Israeli Army, which had been almost exclusively focused on Palestinian terrorist threats for many years. Indeed, many Israeli Army casualties were the result of Hezbollah's antitank guided missiles, weapons that had not confronted Israel since the 1980s and that constituted a threat for which Israeli units were largely unprepared.[7]

Thus, although analysts have urged the United States to pay heed to the lessons the IDF has learned from its ongoing operations in the West Bank and Gaza against nonstate threats,[8] the Second Lebanon War showed that these threats can present a qualitative difference that requires fundamental adaptations in training and doctrine.

As the U.S. military has learned in Iraq and Afghanistan, nonstate adversaries, despite being labeled "low-intensity threats," can be quite effective and difficult to handle. In addition to low-intensity threats, the U.S. military must also prepare to confront state adversaries armed with nuclear weapons and near-peer competitors armed with a wide array of capabilities. Like Israel, the United States must be prepared to operate effectively against this range of threats. Thus, Israel's recent experience in dealing with both an insurgency in the Palestinian Territories and a well-trained and well-equipped militia in Lebanon—a hybrid threat—while simultaneously maintaining readiness for operations against Iran, Syria, and Lebanon should be instructive for the U.S. military.

---

[6] Sergio Catignani, *Israeli Counter-Insurgency and the Intifadas: Dilemmas of a Conventional Army*, London: Routledge, 2008, p. 192.

[7] Catignani, *Israeli Counter-Insurgency and the Intifadas*, p. 193

[8] Bruce Hoffman, "The Logic of Suicide Terrorism," *The Atlantic Monthly*, Vol. 291, No. 5, June 2003.

This chapter was developed using a number of primary and secondary sources and interviews with IDF personnel, defense analysts, U.S. officials who work on issues related to Israel and the Middle East, and defense analysts in both the United States and Israel. We wish to note that the IDF is, as an organization, notoriously difficult to examine. As one scholar wrote in 2008,

> there are major difficulties in studying the IDF and Israeli security in general. Yoram Peri indicated the crux of such difficulties when writing that: "The all-encompassing nature of war in Israel and the centrality of security to national existence have created a situation whereby numerous spheres . . . fall within the security ambit and are enveloped in secrecy." So ingrained is the secretive mind-set of the Israeli security establishment that native researchers with ties to the IDF have stated that even data on the Israeli reserve army is hard to access or find.[9]

There are, however, a number of documents that examine IDF performance during the Second Lebanon War. The IDF commissioned many internal reviews, as did the Israeli government. The most notable government study was produced by a high-level commission, headed by retired judge Eliyahu Winograd, that focused on decisions made by Israeli leaders during the conflict.[10] This chapter relies on these published reports but also considers other sources of information about the IDF, particularly interviews with IDF officers conducted during our fieldwork in Israel.

---

[9] Catignani, *Israeli Counter-Insurgency and the Intifadas*, p. 13.

[10] For a discussion of the IDF efforts, see Alon Ben-David, "Debriefing Teams Brand IDF Doctrine 'Completely Wrong,'" *Jane's Defence Weekly*, January 3, 2007. The Winograd Commission issued classified interim and final reports, but it also issued summaries to the general public. See "English Summary of the Winograd Commission Final Report," NYTimes. com, January 30, 2008; Israeli Ministry of Foreign Affairs, "Winograd Commission Submits Interim Report," Web page, April 30, 2007.

Strategic Demands and Focus

Strategic Imperatives and Priorities

It is an understatement to refer to Israel's position as precarious. The country has less than 21,000 km² of land, which makes it smaller than the state of New Jersey. At its narrowest, Israel is a mere 10 km wide. Israel is also small in terms of its demography: Its population is approximately 6.5 million people. In comparison, there are 19 million people in Syria and 80 million in Egypt.

Israel is also surrounded by states and nonstate entities that it has fought since its inception. Such conflicts include the war of independence in 1948, an engagement in the Sinai in 1956, the Six-Day War in 1967, a war of attrition with Egypt in 1970–1971, the Yom Kippur War in 1973, the First Lebanon War in 1982, the First Intifada in 1987–1993, the Second Al-Aqsa Intifada in 2000–2005, and, most recently, the Second Lebanon War in 2006. Israel currently enjoys peaceful, though somewhat distant, relations with Jordan and Egypt, but it has a tense relationship with Syria and a fragile one with Lebanon. States that do not share a border with Israel—most notably, Iran—have also expressed hostility toward Israel. In addition to these neighbors, Israel must contend with hostile nonstate adversaries. In the West Bank and Gaza, it faces threats from Hamas, Islamic Jihad, and Fatah; in Lebanon, it faces threats from Hezbollah.

Israel has responded to these often-existential threats in part by building a military that relies on quality rather than quantity. It invests heavily in high-tech weaponry (from tanks to aircraft to nuclear weapons) to deter attacks and to defend itself if deterrence fails. Israel provides personnel for its armed forces through mandatory national service and maintains a reserve force that comprises a significant portion of the country's population. Some have argued that the key to Israeli security lies in maintaining a "qualitative military edge" vis-à-vis its adversaries; Israelis define this "edge" as "the ability to sustain credible military advantage that provides deterrence and, if need be, the ability to rapidly achieve superiority on the battlefield against any foreseeable combination of forces with minimal damage and casualties."[11]

Israeli doctrine reflects the country's precarious security situation and is based on the following key points:

- Israel cannot afford to lose a single war.
- Israel is defensive on the strategic level and has no territorial ambitions.
- Israel desires to avoid war through political means and through a credible deterrent posture.
- Preventing escalation is an imperative.
- Israel desires to determine the outcome of war quickly and decisively.
- Combating terrorism is an imperative.
- Israel desires a very low casualty ratio.[12]

Before summer 2006, Israel seemed to be enjoying a period of relative calm with respect to its security. It faced few immediate threats to its existence. With the fall of Iraqi President Saddam Hussein, Iraq was removed as a threat. In addition, Israeli and other military observers judged that Syria posed less of a conventional threat than it had in the past. With Libya voluntarily eliminating its weapons-of-mass-destruction capability, Israeli politicians saw an opportunity to reduce spending on defense.[13] Israel was facing, and has since continued to face, the threat of suicide attacks and indirect-fire attacks from the West Bank and Gaza, but, at the time, these were relatively low-level threats that did not directly menace the continued existence of the Israeli state. The surprisingly poor performance of the IDF in the Second Lebanon War, however, shocked Israel out of its complacency, and, as we discuss in

---

[11] Quoted in William Wunderle and Andre Briere, "U.S. Foreign Policy and Israel's Qualitative Military Edge," Washington Institute for Near East Policy, Policy Focus No. 80, January 2008, p. 3.

[12] Israel Defense Forces, "Main Doctrine," Web page, undated.

[13] Alon Ben-David, "All Quiet on the Eastern Front, so Israel Will Revise IDF Organization and Doctrine," *Jane's International Defence Review*, March 1, 2004.

this chapter, the country has placed a new emphasis on bolstering its military capabilities.

The Role of Out-of-Country Deployments in National Strategy

The one advantage that a small country situated in the midst of multiple threats enjoys is that there is little requirement to deploy long distances to defend itself. Because Israel's adversaries are close by or are even operating within Israeli borders, there is little need for the IDF to deploy far from home to defend Israeli interests. There are exceptions, however. The IAF's strikes on the Iraqi nuclear reactor at Osirak in June 1981 and on a nuclear reactor in northeastern Syria in September 2007 show the country's capacity for long-range air operations. Additionally, although Israel clearly believes that the IAF must answer the threat posed by Iran (and, to a large extent, Syria), it is also clear that Israel has no intention of deploying ground forces far beyond its borders.

How Israel's Strategic Imperatives Compare with Those of the United States

Israel faces a higher level of threat than the United States does. Although breathless concern over Israel's immediate elimination is likely overdone, it does not take much imagination to come up with a scenario that would threaten the continued existence of the Israeli state. Unlike Israel, the United States is bordered by friendly states and does not face significant threats from within its borders. The United States is also removed from many of its adversaries by thousands of miles of ocean. In addition, the United States is a large country, and it enjoys material abundance and a large population from which to draw a military force.

Israel's small size has led the country to place an emphasis on conducting quick campaigns whenever doing so is possible. Short campaigns offer the hope of limiting damage to Israel's relatively small force and to its populace. Smallness has also contributed to Israel's willingness to engage in preemptive strikes, such as those it undertook against Egypt, Jordan, and Syria in 1967; Iraq in 1981; and Syria again in 2007. Although the U.S. military does not prefer to engage in prolonged campaigns, the strategic depth offered by the United States' relatively isolated international position and vast resources, both in terms of people and raw materials, does make longer engagements more feasible for the United States than for Israel.

The United States enjoys many advantages compared with Israel, but Israel's precarious position does entail some advantages. Israel's vulnerability forces the country to be more focused on specific threats. Its need to mobilize its population as a reserve force raises national awareness of the country's military needs and creates a level of solidarity among Israelis that would be foreign to most Americans. In addition, the imminence of Israel's threats has led the country to try to gain and maintain deep knowledge about its adversaries—who they are, how they think, and how they act.

Despite the differences between the U.S. and Israeli strategic imperatives, there are significant similarities that bear notice. For example, both countries have struggled—and will continue to struggle—to balance between preparing for different types of military threats. The small size of Israel's population, coupled with the country's smaller pool of other resources, has forced Israeli defense planners to make difficult choices about how to prepare for different types of challenges. Israeli choices, and the consequences of these choices, are instructive for U.S. defense planners.

The Defense Establishment

Israeli defense has experienced a number of changes in recent years. The lack of a major conflict involving Israel since its intervention into Lebanon in the 1980s, the toppling of President Hussein, U.S. forces in Iraq, perceptions of a reduced Syrian threat, and fiscal concerns led Israel to cut spending on defense beginning in 2003. In June 2003, for example, Israel announced the Kela (Catapult) 2008 program, which sought to reduce the number of armored units by 25 percent and the number of reserve brigades by 30 percent. It also called for a 10-percent cut in personnel across the IDF. The IDF announced that, because of these cuts, reserve units would only be able to engage in full exercises during one out of every three years, and they cut reserve training to two periods of four weeks each year.[14] Furthermore, Defense Minister Shaul Mofaz sought to reduce military conscription and shorten reserve duty.[15] Although ground forces were cut significantly, air and naval forces were more or less spared.[16]

In addition to these changes in equipment, manning, and training, the IDF changed its doctrine and its thinking about how to fight. In 1995, the IDF established the Operational Theory Research Institute (OTRI). OTRI was initially headed by retired Brigadier General Shimon Naveh, whose work became well-known across the Israeli and U.S. defense communities. General Naveh created an approach to military operations that he called *systemic operational design* (SOD).[17] During his time at OTRI and afterward, General Naveh promoted the use of SOD as a way to conceive of and plan military operations. Proponents of SOD describe it as being based in philosophy and science.[18] SOD uses terms more often associated with French literary theory than military operations, and critics have often accused it of being unnecessarily complex.[19] This chapter is not the place for a detailed discussion or critique of SOD or General Naveh, but we would be remiss not to point out that both have had an impact—negative, according to some—on IDF thinking and operational planning.[20]

---

[14] Ben-David, "All Quiet on the Eastern Front."

[15] Efraim Inbar, "How Israel Bungled the Second Lebanon War," *Middle East Quarterly*, July 1, 2007.

[16] Anshel Pfeffer, "The Defense Establishment's Financial Brinkmanship," *Jerusalem Post*, August 28, 2006, p. 3.

[17] Matt M. Matthews, *We Were Caught Unprepared: The 2006 Hezbollah-Israeli War*, The Long War Series, Occasional Paper 26, Fort Leavenworth, Kan.: Combat Studies Institute Press, 2008, pp. 24–28. See also Shimon Naveh, *In Pursuit of Military Excellence: The Evolution of Operational Theory*, London: Frank Cass, 1997.

[18] Tim Challans, "Emerging Doctrine and the Ethics of Warfare," paper presented at the "Joint Services Conference on Professional Ethics," Fort Leavenworth, Kan., 2006.

[19] Matthews, *We Were Caught Unprepared*, pp. 24–28.

[20] Yotam Feldman, "Israeli Army Recuperates Radical Theory," *Ha'aretz* (Israel), October 26, 2007. It is also important to note that General Naveh, who was, at the time of writing, a consultant for the U.S. Army at Fort Leavenworth, has had a marked influence on USMC and U.S. Army conceptions of operational art.

In April 2006, just a few months prior to the Second Lebanon War, the IDF produced a new basic doctrinal document. This document was probably released too shortly before the war to have made, on its own, a significant impact on IDF operations in Lebanon. The document is, therefore, better thought of as a reflection of IDF thinking at the time rather than as a cause of IDF behavior in the Second Lebanon War. Although the document is classified, all reports characterize the ground force component as being focused on low-intensity conflict (LIC). It contains an amalgam of concepts relating to effects-based operations and to General Naveh's SOD. It also calls for relying on standoff fires, delivered primarily from the air, as the principal means of prosecuting wars.[21]

The IDF underwent a period of intense self-scrutiny after its disappointing performance in Lebanon. It conducted some 50 internal reviews and underwent a high-profile examination by a commission headed by former acting Israeli Supreme Court Judge Eliyahu Winograd. The Winograd Commission's interim report argued that flaws in the IDF's training, operational doctrine, and organization contributed to the outcome of the war.[22] The commission's final report found the ground forces to be insufficiently prepared and charged IDF leaders with holding "a baseless hope that the capabilities of the air force could prove decisive in the war."[23] Other postmortems found IDF doctrine and orders obtuse and difficult to understand.[24]

The IDF responded to these and other critiques by going back to basics. In January 2007, they placed the cuts from Kela 2008 on hold, and, in September 2007, the Israeli government announced a new defense plan, Teffen 2012. This plan calls for a new emphasis on building up IDF ground forces, including the creation of new infantry brigades. It also foresees adding "hundreds" of Namer heavy infantry fighting vehicles, several dozen Merkava IV main battle tanks, and a

---

[21] Discussions with IDF officers, Tel Aviv, March 2–5, 2008.

[22] Israeli Ministry of Foreign Affairs, "Winograd Commission Submits Interim Report."

[23] "English Summary of the Winograd Commission Final Report."

[24] Siboni, "The Military Campaign in Lebanon," p. 68.

number of tactical UAVs for use at the battalion level.[25] Israel also put new emphasis on training with its decision to make the training budget for 2007 double that of 2006.[26] There have also been doctrinal reforms. IDF training, particularly in the Israeli Army, has gone back to basics and is focusing on bedrock combined-arms fire-and-maneuver tactics and skills, using such terms as *attack* and *defense*.[27] Additionally, there has been greater cooperation between the Israeli Army and the IAF in the realms of intelligence, surveillance, and reconnaissance; the integration of UAVs; and close air support. Indeed, the IAF is returning tactical air-control capabilities—which had been removed in the years before the Second Lebanon War—to Israeli Army brigades.[28]

The Components

The IDF is governed by the Ministry of Defense, but, in practice, the ministry is relatively weak and has a small staff. The IDF has a General Staff that is better resourced for oversight. In addition to the General Staff, the IDF is organized into a mixture of medium-based and territorially based entities. There are four territorial commands (Northern Command, Central Command, Southern Command, and Homeland Command). The IAF and the Israeli Navy are separate services that have both operational and management responsibilities. That is, they exercise C2 over fielded forces and generate forces by training, organizing, and equipping them. An Israeli Army force command for ground forces is primarily responsible for training, organizing, and equipping, but it does not direct forces in combat. The territorial commands share training, organizing, and equipping responsibilities for ground forces with the Israeli Army force command, but IDF policies of the early 2000s passed most of these functions to the Israeli Army force command. Since the 2006 Lebanon conflict, however, these responsibilities

---

[25] Jane's World Armies, "Israel," Web page, date not available.

[26] Yaakov Katz, "IDF Readying for Gaza Incursion—But Not Yet," *Jerusalem Post*, September 6, 2007, p. 3.

[27] Discussions with IDF officers, Tel Aviv, March 2–5, 2008.

[28] Discussions with IDF officers, Tel Aviv, February 8–19, 2009.

have once again been shared between the territorial commands and the Israeli Army force command.

One aspect of Israeli C2 that has not changed is the separate chains of command for ground, air, and naval forces. It is unclear to many observers exactly how forces from the three mediums coordinate with one another in combat. The extent of the authority of the territorial commands (compared to that of the General Staff) in terms of directing operations is also less than clear and has likely varied over time.

Human Resources

As previously noted, Israel relies on conscription to fill much of its requirement for active-duty manpower. The IDF's male officers serve for 48 months, enlisted male personnel serve for 36 months, and females serve for 24 months.[29] A plan to limit enlisted-male service to two years was scrapped in the aftermath of the Second Lebanon War.[30]

Although conscription and universal service are becoming more unusual in Western-style democracies, Israel's extreme vulnerability makes it unlikely that the country will consider moving to a volunteer-based force. In addition, conscription brings with it the benefit of instilling more of a sense of national purpose and national identity than would otherwise be the case. Nevertheless, there are signs that Israel has made some moves toward professionalizing the IDF.[31] For example, the IDF has recently instituted a two-year company commander's course to teach leadership and other skills to its best officers early in their careers.

As previously noted, Israel is highly dependent on its reserves. It maintains a force of 565,500 reservists, and only 176,500 personnel are on active duty.[32] Calling up the reservists is no small matter. The IDF

---

[29] Richard Weitz, *The Reserve Policies of Nations: A Comparative Analysis*, Carlisle, Pa.: Strategic Studies Institute, 2007, pp. 97–98.

[30] Alon Ben-David, "IDF Shifts Focus to Ground Forces," *Jane's Defence Weekly*, January 10, 2007.

[31] Stuart A. Cohen, "The Israel Defense Forces (IDF): From a 'People's Army' to a 'Professional Military'—Causes and Implications," *Armed Forces and Society*, Vol. 21, No. 2, Winter 1995, pp. 237–254.

[32] International Institute for Strategic Studies, *The Military Balance 2008*, p. 246.

pays reservists based on their civilian salaries, which can be somewhat large in the case of Israel's high-tech workforce. Furthermore, calling up reservists imposes costs on the Israeli economy by removing laborers from the workforce. In a country of only 6.5 million citizens, removing thousands of workers can have a significant impact.

The limitations associated with the use of reservists make it untenable for Israel to engage in a long-term occupation of foreign territory. The country cannot afford to keep a large number of reservists mobilized for an indefinite period of time. This consideration contributed to Israel's decision to withdraw forces from Lebanon in 2000.[33]

Another feature of reliance on reservists is that, compared with active-duty troops, reservists lack readiness. Because reservists are not constantly practicing the military art, it is unlikely, perhaps impossible, that they will be as prepared for operations as their active-duty counterparts.

Observers of the IDF report that the reservists serve a useful function due to their extreme candor. Many reservists are motivated primarily by their sense of patriotism and a desire to serve their country, and many are unconcerned about advancing their careers. In after-action reviews of the Second Lebanon War, for example, reserve officers issued vehement critiques of the IDF's performance. Protests by reservists in the aftermath of the Second Lebanon War also played a role in the establishment of the Winograd Commission.[34]

Statutory Considerations

Although Israel will continue to rely on mandatory military service, there are also signs that conscription is becoming less than universal for Israeli youth. One observer estimates that almost 25 percent of Israelis do not serve due to religious, medical, and other exemptions.[35]

Israel has attempted to legislate incentives for employers to hire reservists, but it has experienced difficulty in enforcing sanctions against employers who fire reservists for absenteeism. Enlisted reserv-

---

[33] Weitz, *The Reserve Policies of Nations*, p. 103.

[34] Joshua Brilliant, "Analysis: Reservists Demand War Probe," UPI.com, August 21, 2006.

[35] Discussions with U.S. military attachés at the U.S. Embassy, Tel Aviv, March 7, 2008.

ists can be called up for a maximum of 36 days each year, but IDF officials are seeking to limit reserve days to 54 days over three years for enlisted ranks and 84 days per year for officers. But, these limits, if they are implemented, will not be in place until 2011 at the earliest.[36] Furthermore, since the Second Lebanon War, reservists in key positions have been spending more time than was usual before on active duty to maintain proficiency.[37]

Manning Strategies and Recruiting and Retention Considerations

As previously discussed, Israel relies on a combination of active and reserve forces. The principal source of manpower is conscripted forces, and service is, with some exceptions, universal. Furthermore, there are no service academies or ROTC-like commissioning programs; instead, potential officers are identified early in their compulsory service and are given additional training. Some become regular, active-duty officers after serving as conscripts and serve on contracts after their initial four-year term of service. Our discussions with Israeli officers and U.S. Embassy Tel Aviv attachés indicated that, given the opportunities that exist in the private sector, it is difficult to retain talented junior officers. Nevertheless, the greatest difference between the IDF and the U.S. armed forces is the absence in Israel of a long-serving NCO corps in the active component; in Israel, NCOs leave active duty after the end of their period of mandatory service.

The Israeli officers we interviewed said that they had attempted to create an NCO corps in the regular army but that doing so was too expensive.[38] Consequently, the enlisted/NCO turnover in the IDF is essentially 100 percent every three years. It is our sense that this reality forces officers in the regular forces to do tasks, both administrative and warfighting, that are the responsibility of mid- or senior-grade NCOs in the U.S. armed forces.

---

[36] Amir Kidon, "Chief Reservist Officer Speaks," Israel Defense Forces Web site, May 7, 2007.

[37] Discussions with Israeli Army officers, Tel Aviv, March 2–5, 2008.

[38] Discussions with IDF officers, Tel Aviv, March 2–5, 2008.

Priorities: The Mission Set and the Range-of-Operations Focus

As previously noted, Israeli defense planners talk about facing a rainbow of operations, a concept similar to the range of military operations discussed by U.S. defense planners. During the period between Israel's foray into Lebanon in 1982 and the 2006 conflict in the same country, the IDF tended to focus on LIC. It concentrated on thwarting Islamic radicals who operated primarily in Lebanon (until 2000), the West Bank, and Gaza. Accordingly, IDF training focused on tasks associated with LIC. During this period, the IDF cut training on antitank tactics, long-range reconnaissance, and the use of mortars.[39]

Given the perception prior to 2006 that Israel faced a less-threatening security environment, the Israeli reserves were largely used as fillers for active forces executing LIC. Consequently, the reserves' overall collective training and readiness suffered significantly, particularly for high-intensity conflict (HIC). A 2008 report by Israel's Institute for National Security Studies shows the precipitous decline in Israeli Army reserve capabilities over the years:

> Between 1990 and 2004 the length of annual reserve duty was cut by 75 percent, from ten million days a year to two and a half million days in 2004. Routine security tasks were transferred to standing units, and training exercises were stopped almost entirely. The major exception of these years was 2002. The ready use of reservists in Operation Defensive Shield (April 2002) led to a sharp increase in investment in training in that year, and to a focused change in awareness. Yet the result of that unscheduled investment in the campaign and its ramifications (despite its clear success) was a decision in 2003 to stop all training of reservists. Finally, in 2005 the new reserve military service bill, which proposed limiting service to fourteen days a year, passed its first parliamentary reading and the discharge age was lowered to forty. Overall, the general trend in the IDF up to the Second Lebanon War was a reduction in the size of the combat forces and an ongo-

---

[39] Discussions with IDF officers, Tel Aviv, March 2–5, 2008.

ing examination of the possibility of lowering the discharge age and exempting civilians from reserve duty.

So as to avoid losing all of the manpower, the meager usage of reservists notwithstanding, the IDF decided that reservists would be viewed as a reserve pool and if the need arose for such reserve forces, they would be mobilized, equipped with equipment stored in emergency storage facilities, trained quickly, and dispatched to the battlefield. On July 12, 2006 the Second Lebanon War broke out and it was decided to call up three divisions of reservists. The reservists arrived on the battlefield after a long period without training, without suitable equipment, and with very little knowledge of the missions and capabilities.[40]

The same article also highlights how the reserves were used in LIC:

The reserve forces, especially those involved in maneuvers (armored corps and artillery) were detached from their particular field of warfare for many years. Operational activity in the territories required only specific abilities. For example, tank personnel engaged in arrests instead of conducting tank maneuvers, and infantry personnel were assigned to checkpoints and fighting in urban areas in small fighting teams, instead of classic warfare practice, such as advancing and taking positions as part of regimental and divisional warfare. The rationale was as follows: since "the next war" is not "supposed" to involve the large scale use of reservists, and as reservists can "always" be trained if a significant war breaks out, training of reservists disappeared from the IDF's multi-year training programs. The only training that occurred was preparation for specific missions. In other words, no classic warfare needs were addressed.[41]

---

[40] Yoaz Hendel, "The Reserves Comeback," *Strategic Assessment*, Vol. 10, No. 4, February 2008, pp. 37–38.

[41] Hendel, "The Reserves Comeback," p. 38.

As a result of the focus on LIC, units had less knowledge of major combat than would otherwise have been the case. One of the common critiques of the IDF's operations in Lebanon in 2006 was that the Israelis had become so focused on the low-intensity operations in Gaza and the West Bank that they were ill-prepared for the different sort of threat posed by Hezbollah. During the conflict, there were reports of entire Israeli units stopping operations while under fire to assist fallen comrades. Although this might be appropriate behavior during a COIN campaign, it placed the units at risk when they were under heavy fire. There were also reports that units were unfamiliar with how to use mortars, tanks, heavy machine guns, and other weapons more often associated with HIC than with LIC.[42]

It is unclear how an adversary like Hezbollah fits into the HIC-LIC paradigm. Hezbollah is a nonstate adversary, which seems to indicate that operations against it would fall under the LIC paradigm. On the other hand, Hezbollah's use of sophisticated weapons and particular tactics resembled challenges faced during HIC. This mixture of threat characteristics is what led many to label Hezbollah a "hybrid" threat.[43] IDF briefers describe Hezbollah as a "nonstate actor with state capabilities."[44] The organization has UAVs, thousands of trained fighters, and thousands of rockets.[45] In the aftermath of the Second Lebanon War, however, IDF training has been much more focused on HIC training because Israel now believes combined-arms fire and maneuver competencies are required against such adversaries as Hezbollah.[46]

The Missions of the Components

The Israeli Army is widely considered to be the dominant service in the IDF. The IDF has only had one chief of staff from the IAF, Lieutenant

---

[42] Discussions with IDF officers, Tel Aviv, March 2–5, 2008. See also Catignani, *Israeli Counter-Insurgency and the Intifadas.*

[43] See, for example, Frank G. Hoffman, "Hybrid Warfare and Challenges," *Joint Force Quarterly*, No. 52, 1st Quarter 2009, pp. 34–39.

[44] Discussions with IDF officers, Tel Aviv, March 2–5, 2008.

[45] Discussions with IDF officers, Tel Aviv, March 2–5, 2008.

[46] Discussions with IDF officers, Tel Aviv, March 2–5, 2008, and February 9–19, 2009.

General Dan Halutz. His association with the failures of the Second Lebanon War (and his subsequent resignation) make it unlikely that there will be another non–Israeli Army officer in that post in the foreseeable future. In addition to dominating the General Staff, the Israeli Army dominates the territorial commands and considers itself to be the supported service. The dispute between Israel and its Arab neighbors is territorial, and, in the conflicts between them, ground forces have played the decisive role.[47]

The IAF, which operates almost all IDF aircraft, both fixed and rotary wing, provides fires to support the Israeli Army, but its primary roles are to defeat enemy air forces and launch long-range attacks against such adversaries as Iraq, Syria, and Iran. Since 2003 in particular, it has also played a role in operations against nonstate adversaries, but its focus has, traditionally, been on HIC.

The Israeli Navy is a small force that focuses mostly on patrolling Israeli territorial waters and interdicting weapons being smuggled into Gaza. There has been some debate in Israel about whether the force should function more as a coast guard.[48] The Israeli fleet numbers around 20 ships. It has three submarines and plans to add two more. It is working to incorporate UAVs into its operations, seeks to develop an amphibious capability, and has a special-operations force.[49] During Operation Cast Lead (December 2008–January 2009), the Israeli Navy did provide fire support for Israeli Army units and blockade the Gaza Strip, so its role seems to be expanding.[50]

The Range of Operations as Reflected in Preparations and Focus
As previously mentioned, prior to the Second Lebanon War, the IDF focused on LIC. Many IDF officers and other members of the Israeli defense community believed that the country had moved "beyond

---

[47] Discussions with IDF officers, Tel Aviv, March 2–5, 2008.

[48] Alon Ben-David, "IDF Ponders Navy or Coast Guard Role," *Jane's Defence Weekly*, June 13, 2007.

[49] Discussions with U.S. military attachés at the U.S. Embassy, Tel Aviv, March 7, 2008.

[50] Anthony H. Cordesman, "The 'Gaza War': A Strategic Analysis," final review draft, Center for Strategic and International Studies, February 2, 2009, p. 18.

the era of major war" and thought that there was an opportunity to decrease defense spending and curb efforts to increase IDF readiness for large-scale conflict.[51] Furthermore,

> service in the West bank and Gaza strip became mandatory for career advancement. . . . The situation on the Lebanese border was less auspicious, even for those at the level of staff officers. In general, the best commanders were assigned to the occupied territories, not to Lebanon.[52]

With the IDF focused on countering Islamic radicals in the West, the force was ill-prepared for the type and level of threat posed by Hezbollah in southern Lebanon in 2006. There were failures at each major level of war. At the strategic level, decisionmakers failed both to articulate clear and meaningful goals and to construct a concept for how Israel would prevail in such a conflict. For example, Israeli leaders failed to see that punishing the Lebanese people by attacking infrastructure targets (such as the airport in Beirut) would not decrease popular support for Hezbollah. They also failed to provide clear orders to troops and delayed the mobilization of reserve units.[53]

At the operational level, military planners failed to organize and employ IDF forces effectively. Air strikes against long- and intermediate-range rocket launchers succeeded, but the IAF was unable to either stop Hezbollah from firing short-range rockets or force it to accede to Israeli demands. Thus, there is good reason to believe that Israeli military planners relied too heavily on standoff air power to secure an Israeli victory.[54]

If these failures were not enough, the IDF was surprisingly ineffective at the tactical level. As previously mentioned, Israeli troops were

---

[51] Israeli Ministry of Foreign Affairs, "Winograd Commission Submits Interim Report."

[52] Amos Harel and Avi Issacharoff, *34 Days: Israel, Hezbollah, and the Lebanon War,* New York: Palgrave Macmillan, 2008, p. 63.

[53] Yaakov Katz, "Soldiers Fought Bravely, in the Cause of a 'Mistaken Conception,'" *Jerusalem Post,* January 31, 2008.

[54] Sarah E. Kreps, "The 2006 Lebanon War: Lessons Learned," *Parameters,* Spring 2007, pp. 72–84.

either trained for LIC challenges or were not trained much at all. In Lebanon, the Israelis faced terrain and enemy conditions for which they were not prepared. An Israeli journalist noted that, until the conflict in Lebanon in 2006, "at no stage was an Israeli unit required to face down an enemy force of a size larger than an unskilled infantry squad."[55] Hezbollah, although not ten feet tall, was trained and organized into small units and armed with sophisticated weapons, including antitank guided missiles; rocket-propelled grenades (RPGs), including RPG-29s; rockets; mortars; mines; improvised explosive devices; and man-portable air-defense systems. Hezbollah also occupied prepared defensive positions in Lebanon's difficult hilly terrain and urban areas.

Initially, the IDF tried to decide the issue with standoff air and artillery attacks, but this did not stop Hezbollah's rocket attacks on Israel or result in the return of the Israeli soldiers whose capture had precipitated the war. Eventually, Israeli ground forces entered Lebanon, where they encountered real difficulties.[56] One of the key deficiencies in the IDF was that the Israeli Army, highly conditioned by its LIC experience, was initially confounded by an enemy who, even without large formations, presented a qualitatively high-intensity challenge that required combined-arms fire and maneuver and a different combat mindset than was needed to thwart Palestinian terrorists. One Israeli observer noted that

> prior to the war most of the regular forces were engaged in combating Palestinian terror. When they were transferred to Lebanon, they were unfit to conduct combined forces battles integrating infantry, armored, engineering, artillery forces, and other support forces.[57]

This lack of preparation was particularly evident in the cases of Israeli field artillery and air power, which were used almost exclusively for

---

[55] Harel and Issacharoff, *34 Days*, p. 45.

[56] For in-depth examinations of the 2006 Second Lebanon War, see Harel and Issacharoff, *34 Days*; Matthews, *We Were Caught Unprepared*.

[57] Gabriel Siboni, "The Military Campaign in Lebanon," in Brom and Elran, eds., *The Second Lebanon War*, p. 66.

attacks on preplanned targets and rarely in support of ground maneuver. Finally, the highly centralized C2 that had been effective in confronting the intifadas proved problematic when used against Hezbollah.

Quite simply, during the Second Lebanon War, the IDF was not prepared for ground operations when standoff strikes did not force Hezbollah to meet Israeli demands. Older soldiers with experience in Israel's 1980s-era intervention into Lebanon, which was conducted primarily with reservists, were often the only members of the IDF force who had experience or training in critical tasks, such as calling in close air support or operating mortars or heavy machine guns. In addition, Israeli units were expected to meet objectives without proper intelligence about the nature of the threat that they would face. When they were surprised by Hezbollah's superior preparation, sophisticated tactics, and advanced firepower, IDF units failed to respond sufficiently quickly or effectively to succeed. An article by Gabriel Siboni, a researcher at Israel's Institute for National Security Studies, summed up the situation quite clearly:

> During the fighting with Hizbollah, inadequate professionalism of the forces and commanders in some of the combat units was observed. This was the case for regular as well as reserve units. . . . In some instances, the units lacked both the skills and the necessary organic weapon systems required for this type of fighting. Under these circumstances units found themselves trying to adjust rapidly—often successfully—while engaged in fighting. The professionalism of the reserve troops was not better but for different reasons. It resulted from a years-long process during which the army reserves were neglected. The education and training of the officers were shown to be ineffective. The lack of practical training during reserve duty was evident, as was the lack of cohesion of the units, which had a detrimental effect on their operational capability.[58]

The IDF was not prepared to execute combined-arms warfare, which requires integrating ground maneuver; fires (air and ground);

---

[58] Siboni, "The Military Campaign in Lebanon," p. 66.

and intelligence, surveillance, and reconnaissance through a well-trained and extensively practiced C2 system. Consequently, in Lebanon, the IDF could not fully utilize

> the full range of combat capabilities: armor, infantry, reconnaissance, intelligence, engineering, artillery, standoff fire, electronic warfare, attack helicopters, and fighter bombers, [or] combat transport helicopter[s] for deep operations in the enemy's rear and along the flanks.[59]

These deficiencies were particularly apparent in the fires area. Siboni writes that "the artillery forces fired mostly on pre-planned targets and provided only inadequate close support for the ground forces."[60] Furthermore, in the area of air support, there were clearly divergent views between the Israeli Army and the IAF that made integration difficult:

> One of the most important tools of the fighting force is the capability to use close aerial support. The essence of such support is the ability of the commander to enlist aerial fire against targets that were not pre-planned, in response to a changing operational situation. In practice, the air force approached this subject completely differently and interpreted the concept of close air support as another version of attacks on given ground targets.[61]

An Assessment of Component Roles Against Mission Sets

The Israeli Army
We have already noted the deficiencies of the Israeli Army against Hezbollah. The IDF's ground force has been adept at countering Hamas, Islamic Jihad, and other groups operating in the West Bank and Gaza. Israeli budget cuts led to the cancellation of exercises, to equipment

---

[59] Siboni, "The Military Campaign in Lebanon," p. 66.

[60] Siboni, "The Military Campaign in Lebanon," p. 66.

[61] Siboni, "The Military Campaign in Lebanon," p. 67.

shortages, and to ineffective logistics practices that plagued the IDF's performance in Lebanon in summer 2006. In response to criticism about their performance, the IDF has placed a new emphasis on rehabilitating Israel's ground force. It has increased spending and the level of effort focused on training and equipping ground forces. The IDF purchased tens of thousands of personal armor kits, night-vision goggles, and ammunition.[62] It also increased training significantly. For example, in the Golan Heights, it conducted Israel's first brigade-level exercise in six years by staging a combined-arms event involving armor, artillery, and engineering units.[63]

There is evidence that IDF's performance, and particularly that of the Israeli Army, has improved. In reaction to escalating rocket attacks launched from Gaza, the IDF mounted a limited campaign to occupy northern portions of the territory in late February and early March 2008; it then launched a much broader campaign into Gaza in December 2008. By all accounts, the IDF, and particularly the Israeli Army, seem to have performed with much greater skill than it demonstrated in Lebanon in 2006. Still, it is important to note that the adversary in these latter operations consisted of militants from Hamas or the Islamic Jihad, not the better-trained and better-armed Hezbollah units.

The Israeli Navy

The Israeli Navy is relatively small and does not play as prominent a role in its country's defense as either the Israeli Army or the IAF. Nevertheless, it did earn a measure of blame during the Second Lebanon War when the *Hanit*, Israel's premier missile ship, was struck by a C-802 radar-guided missile that was manufactured in China and upgraded in Iran. The *Hanit*'s Barak antimissile system should have been able to counter the missile strike, but the system was not turned on because Israel mistakenly believed that Hezbollah could not fire sophisticated

---

[62] Ben-David, "IDF Shifts Focus to Ground Forces."

[63] Alon Ben-David, "IDF Resumes Training in the Golan Heights," *Jane's Defence Weekly*, March 2, 2007.

missiles at naval targets.[64] Based on discussions, we believe that the Israeli Navy is largely focused on interdicting weapon shipments into Gaza.

The Israeli Air Force

As previously mentioned, the IAF has traditionally focused on HIC and on operations against states without a shared border with Israel (i.e., Iran and President Hussein's Iraq). It has performed well in gaining air superiority over adversaries' air forces, and it has engaged in operations far beyond Israeli borders. After 2003, it shifted its focus more toward LIC to reduce collateral damage and increase its effectiveness at striking individual combatants and small groups.[65] As previously noted, during the Second Lebanon War, the IAF was able to destroy long- and intermediate-range rocket launchers but was not able to find short-range rockets, which are easy to set up and dismantle quickly. This is a frustrating problem for Israel because most of the rockets that Hezbollah launched in 2006 were deployed in an area of 4 square miles. The problem of interdicting mobile or fleeting targets from the air is not unique to the IAF, however, and is reminiscent of similar difficulties encountered by the U.S. Air Force in Iraq and Kosovo.

Israeli doctrine states that air power always plays a supporting role to ground forces, but IDF conceptions of how air power can contribute to military operations are relatively unsophisticated. It is unclear to many observers exactly how IDF ground troops call in close air support. Indeed, it appears that battalion commanders in Lebanon were the lowest-level IDF commanders who called in air strikes.[66]

At the outset of the Second Lebanon War, the IDF chose to rely solely on rotary-wing platforms to provide close air support. This practice is probably adequate during low-intensity operations against adversaries in Gaza and the West Bank, but it is not as likely to work against larger or more-sophisticated adversaries, such as Hezbollah, or a state-

---

[64] Yaakov Katz, "Navy Officers Reprimanded over 'Hanit' Attack," *Jerusalem Post*, January 2, 2007.

[65] Discussions with IDF officers, Tel Aviv, March 2–5, 2008.

[66] Discussions with U.S. military attachés at the U.S. Embassy, Tel Aviv, March 7, 2008.

based military. Moreover, IDF concept developers do not seem to place much emphasis on the role that air power can play by interdicting adversary ground forces on their own.

Specialty Forces

There are a number of special-force units within the IDF. The Sayeret Matkal, commanded by a colonel and composed of regular soldiers and reservists, conducts reconnaissance and counterterrorist missions for the General Staff. (This is the unit that rescued Israelis taken hostage at Entebbe in 1976.) There are also counterterrorist and COIN units with divisions for both the West Bank and the Gaza Strip: The Egoz Commando unit is focused on operations against Hezbollah, and the Sayeret Duvdevean, whose members often disguise themselves as Palestinians, works in both Gaza and the West Bank. The Sayeret Shaldag is part of the IAF. There are also special engineering units, forces that specialize in urban operations, and canine units.[67]

During the Second Lebanon War, there were reports of Israeli special forces operating deep inside Lebanese territory. A naval special-forces unit confronted a Hezbollah rocket-launching unit operating out of an apartment building in Tyre.[68] The Sayeret Shaldag and the Sayeret Maktal conducted a joint raid in the Bekaa Valley. Since the 2006 conflict, the IDF has implemented a plan to combine the various special-operations units under a single command structure.[69]

Training Regimes

Overall Training Methodology

Israeli Army unit training is similar to training conducted by the U.S. Army and the USMC in that the IDF uses lists of tasks to focus prep-

---

[67] Jane's World Armies, "Israel."

[68] Israeli Ministry of Foreign Affairs, "Summary of IDF Operations Against Hizbullah in Lebanon," Web page, August 5, 2006.

[69] Amir Rapaport, "The Unification Struck at Dawn," *Ma-ariv* (Israel), April 10, 2007, in BBC Monitoring, April 12, 2007.

aration for combat. Battalion commanders choose from a universal list of tasks and train their units to achieve capability in those tasks. Training is geared toward the requirements set in existing operational plans, and commanders report readiness data back to their respective branches.[70]

As a conscript-based organization, the IDF places a good deal of emphasis on basic training. There are three different options for basic training:

- Generalized training
  - is provided to women and men with disabilities
  - supplies orientation skills, including the use of basic weapons
  - lasts for one month.
- Corps training
  - supplies infantry training and indoctrination
  - lasts for three to four months.
- Brigade training
  - follows corps training
  - is the most rigorous and specialized of the three training types
  - is conducted at bases run by the various branches
  - lasts for four to five months.[71]

Basic Training for Conscripts

The IDF's Tironut program provides basic combat skills to new recruits. There are several different types of programs conducted within the overarching Tironut program. Disabled individuals enroll in Rifleman 01, soldiers serving in noncombat-duty positions take Rifleman 02, and Rifleman 03 is mandatory for all combat recruits. Rifleman 02 includes training in the use of M-16 assault rifles, the use of standard IDF communication equipment, first aid, and measures to protect against chemical and biological weapons. Rifleman 03 includes all of the elements of Rifleman 02 plus training on the M-240 machine gun and several types of grenades, basic field navigation and survival, and single and

---

[70] Discussions with IDF officers, Tel Aviv, March 2–5, 2008.

[71] Photius Coutsoukis, "Israel Training," Photius.com Web page, undated.

squad combat maneuvers. Each type of training consists of presented material, and, in many cases, practical application. There are also branch-specific programs, such as Rifleman 05 (for combat engineers) and Rifleman 07 (for infantry). In almost every program, recruits are converted from citizens into soldiers and are held to strict standards of discipline. Soldiers are also taught Israeli military history. Trainers are not allowed to physically or verbally abuse soldiers, so discipline is enforced through the imposition of punishments, such as additional push-ups and running. At the end of Tironut, soldiers swear an oath of allegiance to the IDF and receive berets.[72]

The IDF operates a national training center at Tze'elim in the Negev. It is fully instrumented, and it features a control center and debriefing room. It can host brigade-level exercises and incorporate helicopters into training events. There are also specialized centers for training paratroopers, armor, and infantry and for teaching the conduct of operations in urban environments.[73]

As previously mentioned, the IDF tended to neglect training prior to the 2006 conflict with Hezbollah, and it has since placed renewed emphasis on the need to prepare for HIC. There is a renewed emphasis on in-the-field, live-fire training for both active and reserve units.[74] Prior to 2006, the training time of regular units was cut from half of the year to ten weeks, and reserve units trained rarely, if at all. The brigade-level exercise in the Golan Heights in 2007 represented a new commitment by the IDF to exercise large units and train for combined-arms operations. Israeli Army reserves have perhaps benefited the most from this renewed emphasis of training:

---

[72] "Tironut," Wikipedia, date not available.

[73] Jane's World Armies, "Israel." The urban training facility is particularly impressive. Its large size creates an environment where the urban area cannot be dominated by fires from outside the urban area. Additionally, it has several mosques, structures over five stories tall, and buildings with tunnels. Thus, it replicates, in quality and scale, the urban environments the Israelis expect to encounter in Gaza and Lebanon. Two of the authors of this report, David Johnson and Michael Spirtas, visited Tze'elim in February 2009.

[74] Discussions with IDF officers, Tel Aviv, March 2–5, 2008.

After the war, towards the end of 2006, the IDF devised a systematic training program for all reserve forces. After many years of inactivity the reserve forces began to focus on the armored corps conducting training with live fire. In 2007 most of the land based divisions carried out full exercises, including closing gaps in warfare procedure and basics. The IDF is thus undergoing a process of establishing and training all combat forces and combat support forces and, more important, today there is outside civilian and political control of maintaining preparedness of the reserve forces.[75]

Training for New Officers

As previously noted, the IDF selects officers from a pool of conscripted candidates. These officer candidates attend the Balad 1 officer school. Prior to 1990, there were three courses taught at Balad 1:

- Chir, a six-month course for infantry and paratroop officers
- Agam, a combined-arms course for armored-corps, artillery, engineering, and air-defense cadets
- Besisi, a basic officers' course for officers who will serve in noncombat-duty positions.[76]

Balad 1 underwent significant changes in the 1990s. As a result of court decisions, women were integrated into Balad 1 and even allowed to attend pilot training.[77] The Balad 1 curriculum was reformed in several ways:

- The once-separate combat Chir and Agam courses were combined into a new course, which includes a three-month combined-arms phase that is followed by a branch phase. The course culminates in a combined-arms exercise.

---

[75] Hendel, "The Reserves Comeback," p. 38.

[76] Tamir Libel, "'Follow Me!': The IDF Officer School and the Changing Civil-Military Relations in Israel," briefing presented at "3rd International Conference," Military Psychology Center, Ramat Gan, Israel, February 20, 2009.

[77] Libel, "'Follow Me!'"

- The three-month–long combat-support course has been lengthened, and its level of instruction has increased.[78]

Since May 2003, IDF officer training has been conducted

on the basis of function rather than gender. Both male and female cadets earmarked for combat duty attend the combat officer training course, while future combat support and combat service support cadets are trained in a separate course.[79]

Training for Senior Officers

The IDF offers a broad range of officer training. It has a school of advanced military studies for rising officers at the colonel and brigadier-general levels; the school is somewhat similar to the British Higher Commander and Staff Course. The purpose of the Israeli program is to educate these officers in the strategic and operational levels of war. The training is conducted at the Dado Center for Interdisciplinary Military Studies.[80]

Deployment Training

Most IDF units are territorially based. They train to specific operational plans associated with the territorial command to which they are assigned. Because the IDF does not generally deploy far from Israel's borders, there is little difference between their predeployment training and the training conducted in the absence of a crisis. That said, as detailed above, the IDF has struggled in trying to find a balance between preparing for high- and low-intensity operations.

Joint Structures and Training for Joint Operations

The IDF has serious problems when it comes to jointness. According to doctrine, the IAF is supposed to support the Israeli Army, but it is unclear whether the IDF has standard procedures for calling in close

---

[78] Libel, "'Follow Me!'"

[79] Libel, "'Follow Me!'"

[80] Discussions with IDF officers, Tel Aviv, March 2–5, 2008.

air support. In general, it seems as if both the IDF's lack of recent experience with large-unit operations and its focus on the tactical level have made it difficult for the organization to integrate air and ground forces. The 2007 exercise in the Golan Heights did include both air and ground units, and Israel's renewed emphasis on training has included some efforts to incorporate IAF units using live ammunition in support of ground forces, but, in interviews, IDF officers reported difficulties in bringing air and ground forces together in training events.[81]

Training for Coalition Operations

Unlike many other Western militaries, the IDF does not really expect to fight in coalition with other states. Given Israel's pariah status in the Middle East, it is difficult to envision a scenario in which IDF units would operate in concert with forces from another state. Furthermore, Israel's vulnerability and proximity to its adversaries makes it unlikely that the IDF would be sent far abroad to engage in peacekeeping, peace-enforcement, or humanitarian operations.

That said, the IDF does engage in a good deal of training with the U.S. military, mostly at lower levels of organization. It trains with the USMC on quick-reaction operations, mostly at the platoon level. There are also a number of different bilateral ties between the IDF and the U.S. military; between the Israeli Ministry of Defense and the U.S. Department of Defense; between the IDF General Staff and the U.S. Joint Staff; between the Israeli Ministry of Foreign Affairs and the U.S. State Department; and between the IDF and U.S. European Command, JFCOM, and the U.S. Army Training and Doctrine Command. Israel also sends officers to various professional-military-education billets in the United States.

Training Methodologies for Foreign Militaries

For many of the same reasons that the IDF does not engage in coalition operations, it also does not seek to train foreign militaries. One might argue that it would be in Israel's interests to see the Palestinians create a professional military and an internal-security force that can establish

---

[81] Discussions with IDF officers, Tel Aviv, March 2–5, 2008.

order and combat terrorism in the West Bank and Gaza, but it seems beyond the realm of political possibility for Israel to consider offering such assistance or for the Palestinians to accept it.

Comparison with U.S. Regimes

The vast differences in the security challenges facing the United States and Israel, combined with the differences in size and structure between the militaries of the two countries, make it difficult to compare IDF training to that practiced by the U.S. military. The most important comparison lies in the nations' shared need to prepare for both low- and high-intensity threats. Some members of the U.S. defense-planning community have expressed concern that the United States' current focus on preparing for operations in Iraq and Afghanistan could decrease the U.S. military's effectiveness elsewhere.[82] In addition to preparing for stability and reconstruction operations and COIN, both the U.S. and Israeli militaries need to maintain proficiency in MCO in the case of conflict with such adversaries as China, North Korea, Iran, and a resurgent Russia.

Israel, like the United States, uses advanced technology to great effect. However, adversaries of both the United States and Israel have increasingly adopted asymmetric means, such as insurgency, terrorism, and guerrilla warfare, in an attempt to counter U.S. and Israeli conventional military prowess. Both Israel and the United States will need to use all of the means available to them to meet these challenges. Both must seek the proper balance between preparing for low- and high-intensity threats.

Adaptability Training

Prior to the Second Lebanon War, the IDF did an excellent job of adapting to the security demands presented by the West Bank and Gaza. The IDF had also continued to plan for other high-end contin-

---

[82] John T. Bennett, "Mullen: U.S. Military Needs Larger Slice of GNP to Modernize," *Defense News*, November 28, 2007.

gencies (i.e., Syria and Iran), and their adaptations to the realities of LIC were impressive and, in some ways, mirrored the British experience in Northern Ireland after the 1970s.

Israeli forces moved away from using overwhelming force in the Palestinian areas, preferring instead to employ "'low-signature' operations that were often not only more effective, but also domestically and internationally less controversial due to the stealth and rapidity with which they were carried out."[83] Their focus was on effectively eliminating terrorist cells while avoiding Israeli or Palestinian civilian casualties. Intelligence preparation, training, organizational structures, facilities, and equipment evolved to better deal with the day-to-day LIC challenges that were heavily stressing the IDF and other Israeli government agencies. Israeli forces became quite good at LIC, adapting as the Palestinians evolved.

As previously discussed, the IDF was largely unprepared for the Second Lebanon War. Its capacity to adapt had largely been confined to LIC. And, as Israeli author Yehuda Wegman noted, the mindset needed for LIC was not appropriate to the situation the Israelis ultimately confronted in Lebanon:

> Fighting against irregular forces is characterized by [the] use of very small and highly decentralized forces with strict limitations on opening fire, so as to refrain if possible from harming the uninvolved, and above all, by the justified drive to prevent casualties among one's own forces even at the expense of failure to carry out the mission. In addition, because of the low number of events and the existence of sophisticated command and control systems, in this type of fighting the commanders' control of the forces is complete and detailed, down to the level of the individual soldier, and every movement of the forces is directed and monitored at headquarters via the screens.

---

[83] Catignani, *Israeli Counter-Insurgency and the Intifadas*, pp. 114–115. Catignani's book provides a comprehensive assessment about how the Israelis adapted to the challenges of the intifadas. It also concludes with a chapter on the Second Lebanon War.

By contrast, the type of warfare required against massive maneuvering or defending forces in fighting against regular forces demands totally different force deployment modes. The situation changes suddenly and dramatically, the forces are large compared to those used in the other type of fighting, and the limitations on opening fire are relatively few. However, above all, in warfare of this type *the necessity to fulfill the mission under all circumstances and at any cost* is the mindset that must govern the commanders and is what obligates them, as opposed to fighting against irregular forces, to leave their command centers frequently and to place themselves at the head of their forces. . . .[84]

Wegman believes that there are "two distinct cultures of warfare, and as in any transition between cultures, here too the transition from one to the other is problematic and complex, and above all, requires an awareness that such a transition is necessary to begin with."[85] He asserts that the Second Lebanon War

proved that a quick transition to a culture of "the mission above all" is almost impossible for those who have been trained to operate only within the context of the culture of "zero casualties to our forces" that characterizes, and justifiably so, the logic of fighting against irregular forces.[86]

Wegman also believes that Israel's almost exclusive focus on irregular warfare (i.e., LIC) "caused a drastic reduction in training in general, and a total halt to the kind of training that was necessary to prepare for the fighting against Hizbollah in 2006 in particular."[87] Thus, when Israeli Army commanders "were forced to deal with the battlefield reality that prevailed during the Second Lebanon War, they were left without the ability to give it the proper response."[88]

---

[84] Yehuda Wegman, "The Struggle for Situation Awareness in the IDF," *Strategic Assessment*, Vol. 10, No. 4, February 2008, pp. 23–24.

[85] Wegman, "The Struggle for Situation Awareness in the IDF," pp. 23–24.

[86] Wegman, "The Struggle for Situation Awareness in the IDF," pp. 23–24.

[87] Wegman, "The Struggle for Situation Awareness in the IDF," pp. 23–24.

[88] Wegman, "The Struggle for Situation Awareness in the IDF," pp. 23–24.

The IDF is rather adaptable as an institution. Israel has a tradition of self-criticism of the type expected in an open society with a tradition of free expression. The Winograd Commission is not the first high-profile panel to critically examine IDF operations. For example, the Agranat Commission was created to look into the IDF's shortcomings during the Yom Kippur War.[89] In addition to these high-profile panels, the IDF conducts internal reviews, such as the 50 internal studies conducted after the Second Lebanon War. Observers of the IDF report that the organization's use of lessons learned is not perfect or automatic but that the IDF is more responsive to change than the U.S. military tends to be. Clearly, Israel's inherent vulnerability is a strong impetus for the IDF to continuously adapt to Israel's complex security environment.

Key Insights

Some observers of the IDF argue that the force tends to focus primarily at the tactical level and has not developed to a great extent at the operational and strategic levels.[90] The IDF's focus on tactics is understandable given both the immediate nature of the threat that Israel faces and the relatively small scale of the conflicts the country has engaged in, particularly over the past 30 years. The Israeli military values *bitsuism* [performance orientation], and, although pragmatism is a valuable trait in a military, too much emphasis on *bitsuism* can foster anti-intellectualism and an inability to derive larger lessons from discrete data points. It can also make it difficult for an institution to change.[91]

One major lesson from the IDF's experience is that it is inherently dangerous to focus too narrowly on just one type of threat. After the demise in 2003 of President Hussein's regime in Iraq, the IDF and the Israeli military establishment in general emphasized operations against

---

[89] Martin van Creveld, *The Sword and the Olive: A Critical History of the Israeli Defense Force*, New York: Public Affairs, 2002, p. 246.

[90] For example, see Catignani, *Israeli Counter-Insurgency and the Intifadas*, p. 9.

[91] Catignani, *Israeli Counter-Insurgency and the Intifadas*, p. 183.

Palestinian terrorists in the West Bank and Gaza, deeming those threats to be the most-pressing challenges to Israeli security.

Clearly, the Second Lebanon War was a wake-up call for the IDF. In the aftermath of that war, the IDF shifted its focus more toward HIC. This shift did not necessarily involve preparing for MCO against other states; rather, it involved a realization on the part of the Israelis that LIC and HIC are different in both substance and scale. After the Second Lebanon War, the IDF recognized that it must prepare for both LIC and HIC.

The first test of the results of the IDF's reforms after the Second Lebanon War came during Operation Cast Lead in Gaza. It is beyond the scope of this monograph to examine this operation in any detail, however, because writing was completed before Cast Lead. Nevertheless, suffice it to say that the IDF views its performance in Cast Lead as successful and the performance of the IDF as much improved since the Second Lebanon War. Much of the military success during Cast Lead can be attributed to the already discussed return to an emphasis on "basics"—vastly improved (and understandable) planning, training, and integration of air, ground, naval, and ISR capabilities. The ability to make these improvements, however, was fundamentally nested in an important conceptual realization after the Second Lebanon War: Precision, stand-off fires are critical, but not sufficient, to cope with irregular-warfare opponents, particularly if those opponents are operating "among the people." In short, the IDF realized that hybrid opponents, such as Hezbollah and Hamas, require a joint, combined-arms approach that enables integrated fire and maneuver, particularly in complex terrain and in military operations among the people. Although Hamas is qualitatively not as significant a challenge as Hezbollah, Cast Lead showed that the IDF is much more prepared for future hybrid warfare challenges than it was in 2006.[92]

---

[92] Discussions with IDF officers, Tel Aviv, February 10–12, 2009; Washington, D.C., February 26, 2009; Tel Aviv, September 2–10, 2009.

CHAPTER SEVEN

Conclusions

I'm worried that we're losing the edge on our ability to conduct full-spectrum operations and major combat operations. . . . Some people say that we're so busy all we can do is focus on COIN operations and we have no time to focus on major combat operations and I think that is wrong.

—Lieutenant General Rick Lynch, Commanding General,
III Armored Corps[1]

This concluding chapter discusses what we learned from our assessments of the armed forces of China, France, the UK, India, and Israel. We begin with several overarching generalizations and then present specific observations in the areas of training, adaptability, and TAA.

General Observations

Not surprisingly, the training and organizing approaches of the armed forces of China, France, the UK, India, and Israel reflect the demands placed on them by their specific strategic environments. This chapter focuses on identifying areas in which these countries employ different approaches to readiness and operational issues that may offer potential benefits to the U.S. system.

---

[1] Quoted in Kate Brannen, "Ft. Hood Commander Concerned Army Is Losing Full-Spectrum Capabilities," InsideDefense.Com, April 9, 2009.

One principal conclusion must be stated at the outset: The U.S. training system is the envy of the countries we examined, and they attempt to emulate many U.S. best practices, such as CTCs. Additionally, as our research proceeded, it became obvious that the differences between how the United States and other nations train their ground forces are much greater than the differences between how they train their air and naval forces. The ways in which air and naval operations are conducted are much less affected by changes in the geographic and sociopolitical setting than are ground-force operations. Therefore, there is less scope for differences in how countries approach the problem of how to prepare air and naval forces for different contingencies. The principal adaptations required of air and naval forces are those dictated by (1) the relative capabilities of an adversary and (2) the specific ROE imposed by the national command authorities. The adaptations of air or naval forces required by changes in the physical environment and the sociopolitical milieu are much less demanding than those required of ground forces. This is not to say that the former set of adaptations is not demanding: We only wish to note that U.S. air and naval forces face challenges comparable to those faced by the other nations and that all those forces train and prepare for the challenges in similar ways. The major difference is that the U.S. naval and air forces are much larger and their training is generally better resourced. The principal area of commonality shared by all the air forces we examined is the difficulty of integrating those air forces with ground forces. This is partly an issue of insufficient interservice cooperation and different service perspectives, but it is also one of a lack of sufficient meaningful training in peacetime. This was particularly evident in both the performance of the IDF during the 2006 Second Lebanon War and the difficulties faced by Indian forces in Kargil in 1999.

Ground forces face a very different situation than do naval and air forces. All of the states we examined, with the exception of China, are or have recently been engaged in active military operations that range from participation in large-scale combat operations to COIN to peacekeeping to TAA missions. These very different types of operations, in our view, suggest that ground forces face the greatest demands in preparing for multiple types of military challenges. Furthermore, several of the nations we examined take different approaches than does the United States, and, therefore, those cases yield the majority of our insights.

Strategic Imperatives, Range of Military Operations, Specialty Forces, and Human Capital

As previously noted, each of the states we assessed organizes and maintains its military forces to address what it perceives as its strategic circumstances. Not surprisingly, the militaries of France and the UK look most like the U.S. military. Neither nation faces any internal threats that require a military response, and, thus, their militaries are used abroad to pursue national policies and priorities. Both militaries deploy their forces overseas, but these deployments are limited to fit the size of the country's force and budget. Furthermore, deployed French and British forces often serve in a supporting role (e.g., contributing to coalition operations in Iraq or Afghanistan). Both France and the UK also employ significant TAA missions to extend their influence.

China and India, on the other hand, are focused on external threats and internal issues. They participate in few deployments, and those in which they do participate are almost exclusively noncombat operations conducted under the auspices of the UN.

Finally, Israel faces a strategic circumstance that requires its armed forces to prepare for a mix of internal and external threats. Furthermore, these threats demand forces that are trained, organized, and equipped for high- and low-intensity operations and for contending with a state that does not share a border with Israel (i.e., Iran).

The militaries of the states we assessed are generally organized around general-purpose forces designed principally for combat operations. The UK, France, and Israel each visualize a range of operations that their forces may have to execute. Although they also rely on general-purpose forces, China and India mostly prepare their forces for a specific activity (e.g., COIN in India) or for operations relevant to a specific military challenge (e.g., a Taiwan contingency in China). Additionally, China, France, India, and Israel employ paramilitary specialty forces used for internal-security missions that lie somewhere between policing and military action (e.g., COIN, civil support, humanitarian assistance), and China has used its paramilitary forces to support a UN peacekeeping mission in Haiti. However, there does not appear to be a joint culture in any of these nations, except the UK.

The nations we examined (except Israel and, to a degree, China) are replacing conscripts with volunteers, which results in higher costs but increased professionalism and a more-sophisticated operational capability. Israel is the clear exception: There, universal service is still the basis of Israel's active-duty and reserve forces. In the active component, however, the Israeli system has resulted in a military without an NCO corps. Thus, Israeli junior officers pick up duties and responsibilities that, in other militaries, are in the realm of career NCOs.

Insights from Training Approaches

Each of the countries we examined relies on general-purpose forces organized principally for conventional combat operations. In this regard, the five nations are very similar to the United States. There are, however, several differences evident in predeployment training, the use of SMEs, the approach to staff training, the use of CTCs, and adapting to irregular challenges. These differences, described in the following sections, may offer potential best practices for improving U.S. training systems.

Predeployment Training Can Build on Strong Traditional Skills

Training for traditional challenges appears to be highly successful in developing foundational individual and collective skills, skills that the British Army's Land Warfare Center calls the *adaptive foundation*. The term *adaptive foundation* refers to the starting point from which forces can subsequently be adapted to specific operational environments; it is a starting point for adaptability, not adaptability itself. Again, the UK's training and readiness cycle spends most of its time, and all of its CTC resources, preparing forces for traditional challenges. The Sennelager, BGTU, and BATUS CTCs continue to focus on MCO, although they are integrating a more-complex environment into training scenarios.

Preparing units for a specific operational environment (*force generation*) requires a relatively modest commitment of time and resources, provided that units are well-trained in basic military operations (*force preparation*). British forces have earned an enviable reputation at the level of the battle group and below in such diverse theaters as Northern Ireland, the Balkans, Sierra Leone, Afghanistan, and Iraq. OPTAG has the mission of preparing forces for deployment into these various theaters. OPTAG accomplishes this mission with fewer than 200 assigned military personnel (of very high quality). In a training and readiness cycle that lasts 24 months, OPTAG requires around one month to train the trainers; then, it allows the trainers to train their units and conducts a confirmatory exercise. India's various battle schools, including the XVth Corps Battle Schools, the Counter Insurgency and Jungle Warfare School, and the High Altitude Warfare School, prepare units in a similar fashion, represent a similarly modest commitment of resources, and seem to prepare units well for asymmetric challenges.

SMEs Can Provide Crucial Capability

The training of units in the Indian Army is tailored to the specific region and specific operational conditions in which units are stationed. In day-to-day operations, this is a viable approach. When the units deploy to contingencies for which they have not prepared—as in the 1999 war in Kargil, examined in Chapter Five—this approach can prove inadequate. In the Kargil crisis, Indian troops acclimated and trained for tropical COIN operations were not prepared for conventional combat operations in the mountains. The insight from the Indian experience in Kargil is that a small group of SMEs—in this case, mountain-warfare experts—can rapidly infuse capability into units by enabling forces trained for one environment or contingency to improve their performance in a different set of circumstances. In the United States, this SME approach could be a way to improve training for specific deployments or to improve the performance of units that are deployed against contingencies that were not the focus of predeployment preparations. To do this, the U.S. military should take advantage of SMEs for operations across the spectrum and for different types of complex terrain, whether mountainous, urban, or jungle. Furthermore, to leverage these

SMEs, a system of identifying and tracking SMEs across the force has to be in place.

Staff Training Can Serve as a Vehicle to Prepare Forces for Multiple Contingencies

French processes for C2 training offer a potential model for training staffs for multiple types of operational contingencies. The effective transition of French forces in the Côte d'Ivoire in 2004 from peacekeeping operations to irregular warfare demonstrated very agile C2 capabilities. The professional French response undoubtedly owed much to the force's highly unorthodox operational commander, General Poncet, but it also points to the importance of highly trained staffs. French brigade staffs gain their proficiency by conducting three to four times as many CPXs per year as either the U.S. or the British staffs. Furthermore, a number of these CPXs are externally evaluated.

The significantly greater frequency of CPX training in the French force is enabled in part by the fact that the French often train a single echelon at a very reduced scale. Thus, conducting a meaningful CPX does not require coordinating the schedules of multiple HQ at several echelons—an effort whose scope and complexity deter frequent repetition. Technology also plays an enabling role. The French SCIPIO simulation for brigade HQ and above automates many of the entities, reducing the requirements for higher- and lower-control players. Moreover, French operational HQ have relatively few day-to-day oversight and housekeeping responsibilities with regard to their subordinate units. The Commandement de Formation d'Armée de Terre synchronizes unit training calendars directly, and administrative and logistical support comes from the regional commands.

U.S. Army and USMC units that have spent the past several years focused almost exclusively on Iraq, Afghanistan, and COIN have become increasingly proficient in those operations. Nevertheless, there is growing evidence that high-end combat skills are atrophying because of the U.S. military's understandable focus on current wartime challenges. A recent paper by three U.S. Army brigade-combat-team commanders highlighted the urgency of this situation in terms of the fire-support system. They cited Israel's experience in Lebanon in 2006 as a warning:

With each passing month that we continue to let these perishable skills atrophy and lose our expert practitioners, we are mortgaging not only flexibility in today's fight, but our ability to fight the next war as well. This is similar to what happened to the Israeli Defense Forces. Israel's years of COIN-focused operations in the occupied territories cost them dearly in South Lebanon. When the IDF attempted to return to HIC operations, it found itself unable to effectively plan fires, conduct terminal control or deconflict airspace. The IDF's ability to conduct combined arms integration had simply atrophied from neglect. We should consider ourselves fairly warned. We can't afford to lose sight of the critical role artillerymen play in our ability to plan, coordinate, integrate and synchronize our combined arms operation. This is not an artillery branch issue, this is an Army issue, as the Israelis learned . . . the hard way.[2]

The French process of increasing the proficiency of unit HQ seems to be highly effective in enabling units to master transitions, and their ability to do so shows that C2 training yields a high return on a marginal training investment. Thus, directed, evaluated CPXs could provide a training methodology for U.S. forces that could help address concerns, recently voiced by General Casey, about the deterioration of critical integration, synchronization, and other skills required to prevail across the full range of military operations. According to General Casey, "Current operational requirements for forces and insufficient time between deployments require a focus on COIN training and equipping to the detriment of preparedness for the full range of military missions."[3]

The U.S. Army BCTP and other service staff-training programs have long executed rigorous staff training. BCTP has now, very understandably, evolved largely into a mission-rehearsal exercise for deploy-

---

[2]   Sean MacFarland, Michael Shields, and Jeffrey Snow, "White Paper for CSA: The King and I—The Impending Crisis in Field Artillery's Ability to Provide Fire Support to Maneuver Commanders," undated [2008], p. 3.

[3]   U.S. Department of the Army, *2008 Army Posture Statement: A Campaign Quality Army with Joint and Expeditionary Capabilities*, Washington, D.C.: Headquarters, Department of the Army, 2008, p. 6.

ments to Iraq and Afghanistan. We suggest conducting multiple events, perhaps graded, over the course of a year to expose battalion- and higher-level staffs to different types of planning requirements across the range of military operations.[4] Enabling measures for this approach, such as development of simulations similar to the French SCIPIO system, should be considered.

CTCs Can Be Used Differently

Several of our case studies show that other countries believe that their training centers should mainly provide foundational combined-arms fire-and-maneuver training. The militaries build on these skills with predeployment training focused on the specific operational environment to which a given unit is deploying. We believe that reorienting U.S. training to a predeployment model along the lines of OPTAG or the Indian Army's Counter Insurgency and Jungle Warfare School would allow CTCs to return to a principal focus on task-force combined-arms fire-and-maneuver training. We believe that this training is critical to maintaining full-spectrum capabilities and to addressing General Casey's concerns. We are not implying that there should be a return to a "Fulda Gap" model; rather, we believe that the model employed by several of the countries we examined is worthy of close examination by the United States. Additionally, we are not advocating that the CTCs return to portraying a sterile battlefield. BATUS has integrated villages, civilians, and other complications into its training scenarios. Similarly, the Israelis have a sophisticated urban-operations training facility in Tze'elim.

It is logical to assume that when units spend time at CTCs preparing for the operational environment in Iraq and Afghanistan, they are not using that time to train for synchronized brigade and battalion task-force combat operations. It makes eminent sense to prepare units for the specific operational context they will face, but doing so at a CTC sacrifices an opportunity to conduct foundational combined-arms training at facilities uniquely suited to provide this training.

---

[4] U.S. Department of the Army, "Battle Command Training Program," briefing, May 12, 2008.

Contextual training could likely be done elsewhere (at lower cost) or become one component of the CTC experience. Clearly, U.S. CTCs are the locations best prepared to provide combined-arms training in intense simulated combat. And, as the Israeli experience in Lebanon in 2006 shows, intense combat is not so much about scale (i.e., battalion or brigade force-on-force engagements) as about the qualitative challenges hybrid adversaries can pose. Opponents with a modicum of training, organization, and advanced weaponry—like Hezbollah—create tactical and operational dilemmas that demand combined-arms fire and maneuver. Thus, based on their experiences in Lebanon in 2006, the Israelis have reoriented the focus of much of their training—particularly the training conduced at the Tze'elim training center—on HIC. Their subsequent performance in Gaza in December 2008–January 2009 seems to show that this reorientation was wise.

Moreover, because the goal of the training centers in France, the UK, and Israel is foundational rather than finishing, U.S. units might profitably undergo their CTC rotation earlier in their training cycle. The point of maneuver exercises in the UK and Israel is as much to teach staff operations, planning, troop-leading procedures, and basic tactical skills as to teach specific collective tasks. British forces undergo CTC rotations toward the middle of their readiness cycle, and such rotations constitute most of French forces' collective preparations for an operational tour.

Adapting to Irregular Challenges

The sponsor asked that we examine approaches to training forces to adapt to irregular challenges. There is an emerging literature that emphasizes the importance of individual and unit adaptability and, thus, improving methods to train both to be adaptable. Proponents of this training approach argue that

- The United States faces future threats that are irregular and asymmetric.
- Adaptability is key to meeting these challenges.
- Adaptability (and intuition) can be taught.

This adaptation tautology implies that fundamental change across the DOTMLPF spectrum is not necessary to prepare for irregular and asymmetric challenges—well-trained individuals and units can adapt to any circumstance.

Although we generally believe this approach to individuals and units is important, in our view, it is necessary but not sufficient. Our opinion is that it is the role of the institutions within the DoD to prepare U.S. forces for the challenges that they will encounter in specific irregular (and regular) operations. The responsibility for adaptation must also belong to these institutions rather than to individuals and units. This is not to say that teaching critical thinking, decentralizing decisionmaking, and a host of other initiatives are not useful approaches. They are necessary but not sufficient, and they have always been valued, at least in theory, in the past.

That said, the important role of institutions is to provide an appropriate problem-solving framework for use by individuals and units when asymmetries present challenges that existing methods do not address adequately. Perhaps the best recent example of a U.S. military institution adapting itself to new conditions is the U.S. Army's revision of its fundamental concept about how to succeed in war. The 2001 version of Field Manual (FM) 3-0, *Operations*, posited a construct for warfare that had endured in the U.S. Army for nearly 80 years:

> The offense is the decisive form of war. Offensive operations aim
> to destroy or defeat an enemy. Their purpose is to impose US will
> on the enemy and achieve decisive victory.[5]

This was the doctrine that the U.S. Army—a very well-trained and well-equipped force—took into Operation Iraqi Freedom, and, by 2006, it was clear that this approach was not adequate to deal with the insurgency that developed after the end of MCO. Eventually, the U.S. Army revised its approach, publishing, in conjunction with the U.S. Marine Corps, a new COIN manual, FM 3-24/MCWP 3-33.5,

---

[5]   U.S. Department of the Army, FM 3-0, pp. vii, 7-2.

Counterinsurgency Field Manual, that fundamentally changed the basic construct for successful operations, noting that

> the cornerstone of any COIN effort is establishing security for the civilian populace. . . . Soldiers and Marines help establish HN [host nation] institutions that sustain that legal regime, including police forces, court systems and penal facilities.[6]

This institutional adaptation was a precondition for the increasingly successful COIN operations that followed the promulgation of the new doctrine. Quite simply, absent FM 3-24/MCWP 3-33.5, *Counterinsurgency Field Manual*, even the most-adaptable individuals and units were not able to solve the COIN problem across Iraq using FM 3-0, *Operations*.

The United States has been down the road of basing adaptability on the individual many times before (in, for example, Vietnam, Bosnia, Somalia, Haiti, and the Indian Wars). During each of these conflicts, senior leaders believed that the forces they deployed to these conflicts were highly professional and prepared. The U.S. military institutions also valued creative and adaptive leaders. The U.S. Army's 1987 FM 22-103, *Leadership and Command at Senior Levels*, was quite clear in this regard:

> Creativity refers to the ability to find workable, original, and novel solutions to problems. As a skill, it provides senior leaders with the capability to be innovative and adaptive in fast-moving, potentially confusing situations. Its purpose is to find practical solutions to unexpected or tough military problems.
>
> All exceptional leaders and commanders have had a large measure of creative skills. When faced with seemingly impossible problems, they developed solutions which not only worked but turned the situation in their favor. Observers refer to this as "breaking

---

[6]   U.S. Department of the Army and U.S. Marine Corps, FM 3-24/MCWP 3-33.5, *Counterinsurgency Field Manual*, Chicago: The University of Chicago Press, 2007, p. 42.

the code"—the relationship between the situation and pieces of information is seen in new or unusual ways.[7]

Furthermore, FM 22-103, *Leadership and Command at Senior Levels*, emphasized that "developing systematic mental habits and the ability to read critically, think analytically, and communicate effectively is essential for tactical-level and operational-level commanders."[8] In addition, U.S. military staff and war colleges have always prided themselves on teaching their students how to think, not what to think. Admiral Stansfield Turner, a former president of the Naval War College, summed this up nicely:

> War colleges are places to educate the senior officer corps in the larger military and strategic issues that confront America in the late twentieth century. They should educate these officers by a demanding intellectual curriculum to think in wider terms than their busy operational careers have thus far demanded. Above all the war colleges should broaden the intellectual and military horizons of the officers who attend, so that they have a conception of the larger strategic and operational issues that confront our military and our nation.[9]

Thus, it is difficult to recall a time in U.S. history when military leaders did not believe that they were adapting to the realities they faced. Nevertheless, when it becomes clear that the institution's preparations are inadequate to the tasks the military confronts—generally, this is the result of conceptual failures reflected in doctrine—the institution has no option but to rely on adaptive individuals on the ground to sort out the irregularities of the situation while the institution adapts.

This is not to say that the DoD should not continue efforts to increase individual adaptability and understand the true possibilities

---

[7] U.S. Department of the Army, FM 22-103, *Leadership and Command at Senior Levels*, Washington, D.C.: Headquarters, Department of the Army, 2007, pp. 30–31.

[8] U.S. Department of the Army, FM 22-103, p. 84.

[9] Williamson Murray and Richard Hart Sinnreich, eds., *The Past as Prologue: The Importance of History to the Military Profession*, New York: Cambridge University Press, 2006, p. 8.

and limitations of teaching adaptability. At the time of our research, it appeared that the direction the U.S. military was following was based on the assumption that adaptability is a discrete skill that can be taught. Furthermore, it appears that adaptability and intuition are being confused. Much of the work on recognition-primed decisionmaking originates with Gary Klein. Klein studied people working in familiar operational contexts who made decisions that seemed automatic but were indeed highly relevant. At the heart of intuition is pattern recognition. According to Klein,

> intuition is the way we translate our experiences into judgments and decisions. It's the ability to make decisions by using patterns to recognize what's going on in a situation and to recognize the typical action script with which to react. Once experienced intuitive decision makers see the pattern, any decision they have to make is usually obvious.[10]

Thus, our sense is that creating vicarious intuition through training is a highly viable training approach. By exposing leaders and soldiers to different situations via simulation and in predeployment training, one can engrain in them new patterns that will result in intuitive responses to situations that resemble the training scenarios. Of course, to be successful, intuition training must pick the right patterns. Such training also assumes that, under stress, all leaders will recognize the patterns and act as desired. The problem with training individuals to adapt to asymmetrical situations is that the asymmetry may be so great that there are no recognizable patterns. To use one of Klein's examples, such a situation would be similar to placing well-trained firefighters in the middle of a bank robbery and expecting them to adapt. Determining these patterns is the responsibility of the institution as it tries to understand its circumstances when simply doing what is "regular" does not work. Here, again, Klein is useful:

---

[10] Gary Klein, *The Power of Intuition: How to Use Your Gut Feelings to Make Better Decisions at Work*, New York: Doubleday, 2003, p. 23.

> Ultimately, it's important to take a measured approach to intuitive decision making, viewing it as neither an ill advised form of reasoning nor a magical gift. Seek out a balance between intuition and analysis. Both are important sources of power, and both have weaknesses.[11]

Of course, leader development and training programs attempt to replicate the experiences of others (i.e., the lessons learned) and impart them vicariously across the institution. There is a central irony in the way the U.S. military views experience: In the United States, military units are often commanded by officers of a lower grade—and at lower levels of experience—than in other militaries. In several of the other armies we examined, companies are commanded by majors, and brigades are commanded by brigadier generals; in the United States, however, companies are generally commanded by captains, and brigades are commanded by colonels. Additionally, in the case of the French Army, there is a close tracking of officers with prior experience serving in the destination country, and these experienced officers are placed on the staffs of deploying units and specifically sought out to lead training teams.

Training challenges go beyond what to train: They also involve trusting one's training rather than one's previous experiences. In the case of senior leaders, is it reasonable to expect that the trained reaction to the asymmetric pattern will override a career's worth of real experience and training? Under stress, the much more deeply ingrained experiences may dominate the trained intuitive response. Moreover, since 9/11, a generation of soldiers, sailors, airmen, and marines whose only experiences are in irregular warfare has arisen. These personnel are becoming very intuitive after repeated combat tours, and they are gaining a set of experiences that, in many cases, senior officers do not share. Consequently, the U.S. Army and the USMC in particular may be approaching a "clash of intuitions" in which senior officers, who have not gained the deep irregular-warfare combat experience of com-

---

[11] Klein, *The Power of Intuition*, p. 299.

pany and junior field-grade officers, may react very differently to the same asymmetrical situation because of their different past experiences.

There is another inherent risk in trying to teach adaptability to irregular challenges: These challenges become normal. A potential unintended consequence of focusing on preparing units for Iraq and Afghanistan may be that the U.S. military is creating leaders who are deeply experienced in COIN but less prepared for higher-intensity operations. The United States is creating a military whose leaders will, if they entered service after 2004, have had no direct experience in preparing for and executing MCO. As the experience of Israel in Lebanon in 2006 showed, a force deeply experienced and prepared for COIN is not necessarily able to immediately adapt to medium- to high-end combat operations. These high-intensity operations are not a matter of scale but of quality. During "hybrid warfare,"[12] defeating a trained, organized, and well-equipped opponent in good defensive positions, like Hezbollah in 2006, requires combined-arms fire and maneuver by joint forces. This is the central challenge in preparing general-purpose forces for full-spectrum operations. And, for the United States, with its global commitments and responsibilities, the challenge is greater than it is for any of the other countries examined in this monograph.

Clearly, U.S. military institutions understand the imperative to adapt, as shown in FM 3-0, *Operations*:

> Just as the 1976 edition of FM 100-5 began to take the Army from the rice paddies of Vietnam to the battlefield of Western Europe, this edition will take us into the 21st century urban battlefields among the people without losing our capabilities to dominate the higher conventional end of the spectrum of conflict.[13]

Again, we believe that the examples of France and the UK are instructive. These countries rely on systemic adaptation of their forces based on operational lessons learned rather than on seeking to inculcate adaptability in individuals and organizations. Although the

---

[12] For a discussion of hybrid warfare, see Frank G. Hoffman, *Conflict in the 21st Century: The Rise of Hybrid Wars*, Arlington, Va.: Potomac Institute for Policy Studies, 2007.

[13] U.S. Department of the Army, FM 3-0, Preface.

French explicitly prize adaptability, in practice, their training systems have increasingly shifted away from complete reliance on adaptability and toward ongoing adaptation based on operational lessons learned. The British have evolved and perfected this approach over more than 30 years, and the Indian Army continues to extend its use of this method of preparing forces for operations. In none of the countries did we observe collective training efforts expressly focused on inculcating adaptability. We do not mean to imply that no individual elements of the nations' training efforts inculcate adaptability; the French Army's de facto emphasis on commander-leader teams provides one example of such inculcation. However, we did not identify any country for which adaptability, rather than adaptation to a specific operational environment, was the explicit goal of training.

A Comparison of French, British, and U.S. TAA Models

Of all the countries we assessed in this study, it appears that France and the UK have the TAA models that provide insights into improving the U.S. model. Table 7.1 depicts each country's underlying rationale for building-partner-capacity (BPC) programs, approaches to training trainers and advisers, process for selecting trainers (and the effects of such missions on trainers' careers), geographical focus of TAA efforts, and resources devoted to the approach.

All three countries view TAA and BPC as ways in which they can favorably shape and influence the global security environment. That said, their TAA approaches differ significantly in several key areas: trainer selection, mode of deployment, training of the trainers, and career implications for the trainer. We discuss each in turn.

Selection
The U.S. and British processes for selecting trainers and advisers from the conventional forces do not appear to be particularly rigorous, and the assignments are not generally sought out by officers in these two countries. The French model ties the selection process to career

Table 7.1
Comparison of French, British, and U.S. TAA Approaches

Country	BPC Perspective	Training Approach (internal)	Selection of Trainers/Career Outlook	Training Approach (with partners)	Geographical Focus	Resources
France	Expand France's cultural and economic influence. Deploy to countries of operational interest. Integrate with the local population.	Trainers and advisers attend an EMSOME course for between six hours and two weeks. The TAA mission is part of a battalion's normal mission.	Candidates undergo a rigorous interview process with a jury (consisting of a former adviser, a psychologist, and a committee adviser). TAA missions are career-enhancing and lead to command of a battalion. There is no FAO-like program.	The training partner is not the highest priority; rather, the approach is about maintaining French influence. Advisers are embedded in the host nation's ministry of defense. The trainers wear the local uniform.	Francophone Africa The Francophone Caribbean The Balkans Afghanistan	The multiservice RECAMP program Existing regional organizations, particularly in Africa

Table 7.1—Continued

Country	BPC Perspective	Training Approach (internal)	Selection of Trainers/Career Outlook	Training Approach (with partners)	Geographical Focus	Resources
The UK	Enable partners through advisers and trainers.	Large-scale predeployment training conducted at OPTAG lasts for approximately two weeks.	The selection process is not very rigorous.	Advisers are embedded in the host nation's ministry of defense.	East and South Asia	Resources are pooled.
	Change "fair weather friend" reputation.		It is not clear that TAA missions are career-enhancing.	The trainers wear the local uniform.	The Balkans	Priority countries are determined by the Contact Group (the MOD, the FCO, and DFID).
	Integrate with the population.		There is no FAO-like program.	A rigorous needs assessment is conducted at the outset, and training modules are tailored.	The Caribbean	
					The Middle East	
				An MOU is signed with a host-nation senior mentor.	Africa	
				The UK controls training and funding.		

Table 7.1—Continued

Country	BPC Perspective	Training Approach (internal)	Selection of Trainers/Career Outlook	Training Approach (with partners)	Geographical Focus	Resources
The United States	Focus on partners that can help address threats to U.S. interests. Shape and influence partners. Limit integration with the population because of force-protection concerns.	Individual services are in charge of their respective training programs. Discussions about formalizing TAA trainer/adviser training are underway.	The selection process is not rigorous. Missions are not part of mainstream career paths. Missions are viewed as a detriment to promotion by some services.	Advisers are not typically embedded in the host nation's ministry of defense. Advisers do not wear the local uniform. U.S. authorities govern what type of training, equipment, etc. BPC programs can provide.	Global	U.S. government agencies have their own budgets for BPC. Some collaboration occurs on an executing basis.

progression. Advisory duty in the French Army expected from officers who are competitive for advancement.

Deployment

France and the UK have similar TAA models: Advisers are embedded in the partner's ministry of defense and often wear the host-nation uniform.

Embedding advisers has not been the norm for the United States in its vast array of TAA activities with partners. Nevertheless, there are clear exceptions to this general rule. As it did during the Vietnam War, the United States is embedding advisers and training teams in host-nation structures in Iraq, Afghanistan, and the Philippines. Thus, embedding might become the new norm.

Training

The U.S. system for preparing trainers and advisers emphasizes operational and tactical training over cultural training, and what cultural training is offered does not address key points, such as empathy with the advised, covered in the French and British models. Although the French and British predeployment training for advisers lasts only approximately two weeks, in each country, the process appears to do a good job of ensuring that advisers are adequately trained. In the U.S. system, training for TAA lasts between two and six months.

Career Implications

There are no foreign-area officer programs in France and the UK; most of the forces deployed on TAA missions come from the pool of general-purpose forces and are generalists. In the French system, the TAA mission is part of a deployed battalion's normal mission. Furthermore, advisory duty is part of the normal career path, and success in TAA missions is seen as a prerequisite for advancement. This is not the case in the UK or the United States. In the UK, TAA missions are encouraged but not necessarily career-enhancing. In the United States, TAA missions have traditionally not been part of mainstream career paths. Indeed, in the United States, advisory duty has generally been viewed as detrimental to advancement—it was what happened to an officer who was not competitive for more-important, career-enhancing assignments. Clearly, the importance of training the military forces of Iraq and Afghanistan as components of a successful strategy is understood within the U.S. military.

What Should OSD Do About These Insights?

This last section provides overarching insights from our analysis and offers recommendations for OSD to pursue to improve U.S. training practices in four areas: adapting to irregular challenges, preparing the force, defining TAA requirements, and preparing for future challenges.

Adapting to Irregular Challenges

As previously noted, the focus in other countries is on building location-specific intuition as a means of adapting the overall force to the specific contingency. The British and the French also have created the capacity, through OPTAG and EMSOME, respectively, to quickly infuse lessons learned from ongoing operations into the training for those preparing to deploy. This is similar to what the United States is doing at its CTCs. This is done to adapt their militaries to operate in the places to which they are about to deploy and to train individuals within this specific context. This is different from trying to teach adaptability. It is more along the lines of creating deep, vicarious intuition by expanding patterns in training that can be recognized and referred to during operations. The key for the institution is to minimize how long any operational environment remains asymmetric. Thus, our sense is that adaptability is an institutional—not an individual—responsibility. The challenge in training individuals is to prepare them as much as possible for the specific environment of the future deployment.

Nevertheless, there appears to be a need to understand how to identify how individuals respond to complex situations when they are under pressure and when traditional hierarchical chains of command are unavailable to support decisionmaking. This seems particularly important in the case of advisers. Thus, although adaptability may not be a trainable trait, it might be a discriminator for key positions in which the ability to cope with uncertainty is important. That said, our sense is that more empirical investigation is needed before it will be possible to determine whether the recommendations in the IDA study are viable. Our recommendations are as follows:

- OSD should support further empirical research to determine if adaptability can in fact be trained and if an individual's ability to adapt can be determined.
- If adaptability can be assessed and trained, OSD should establish processes to determine which assignments (e.g., advisory assignments) require adaptability.

Preparing the Force

There are several gaps in current processes for preparing the U.S. armed forces for the irregular—and regular—challenges they face. There are multiple populations to prepare. Nevertheless, our sense is that the greatest gap exists at the senior levels. Quite simply, there has never been a deeply substantive or rigorous system of continuing training or education for officers beyond their attendance at a senior-service college at the O-5 or O-6 levels. A number of the nations we examined recognize the need for continuing education beyond that provided by their equivalent of the U.S. senior-service college. The British have a higher-command and staff course, and the Israelis have a course for colonels, brigadier generals, and new division commanders. Because senior U.S. officers are responsible for preparing their units for the challenges of the future and for guiding their training, it seems important to provide them with continuing education. Our recommendation is as follows:

- OSD should assess current programs for the continuing training and education of senior leaders and recommend corrective action.

Defining TAA Requirements

There is currently no enterprise-wide system within the DoD or other U.S. government agencies to identify and prepare American officers for advisory or foreign-military training assignments. These assignments are generally conducted on a one-off basis and are not career-enhancing.

Finally, there is no DoD-wide repository for best practices or lessons learned for these missions. Our recommendations are as follows:

- OSD should work with the Joint Staff and the U.S. military services to set standards for advisers (including selection criteria) and craft directives that ensure that adviser training assignments are career-enhancing. These efforts could be similar to measures taken after Goldwater-Nichols to ensure that joint duty became a viable assignment.
- OSD should create processes to capture and disseminate TAA- and BPC-specific best practices from across the U.S. government and from relevant foreign governments.

Preparing for Future Challenges

One of the central ironies about adapting to and preparing for irregular challenges is that such challenges then become the new "regular" challenges. Israel's performance during the Second Lebanon War is instructive in this regard. After years of adapting to the challenges of the intifadas, the Israeli Army, despite its competence in addressing low-intensity threats, found itself not competent to fight the HIC it encountered in Lebanon. The asymmetry in Lebanon was caused by the inability of the IDF to counter conventional weapons with combined-arms maneuver warfare.

Currently, the U.S. armed forces may be in a condition similar to that of the IDF in 2006. Multiple combat tours in Iraq and Afghanistan have created U.S. units and individuals with deep experience in COIN. Additionally, adapting to the significant demands of the operational environments in these active theaters of war has, not surprisingly, resulted in a diminishment of high-end combat skills among U.S. forces. Thus, the extraordinary proficiency of the U.S. force at doing what it has to do now may in fact be diminishing its capacity—as it did with the IDF—to do something it may have to do in the future. In short, the U.S. military, particularly its ground forces, has lost some of its full-spectrum capability. The concerns of several maneuver commanders about the deterioration of the U.S. Army's fire-support system are instructive in this regard. To understand what needs to be done, it is important to begin with a comprehensive assessment of the state of the deterioration in U.S. military skills and capabilities.

Several of the nations we examined have developed training regimes that assist them in adapting and preparing their units for different operational scenarios. India deployed SMEs to improve unit performance, the French use multiple and evaluated CPXs involving differing scenarios to prepare their HQ, and the British "train the trainer" for several deployment scenarios through their OPTAG process. All of these practices offer promise to improve the current U.S. training system. Our recommendations are as follows:

- OSD should support an analysis to determine which UJTL tasks are atrophying.
- OSD should further assess CPX strategies that train and evaluate HQ for the full spectrum of operations, and it should support the development of exercises that allow staffs to maintain full-spectrum proficiency.
- OSD should assess the potential of SME training and devise processes to identify and track SMEs.

Final Thoughts

During our research, we found that the U.S. military is the source of best practices in many areas in every country we examined. Other militaries recognize that the United States is the only nation that can, at the moment, operate globally and, if need be, independently. Ironically, many of our insights for the U.S. training system are derived from practices that originated in the fact that countries do not posses the military capabilities inherent in the U.S. armed forces. Other nations have had to develop processes to attain full-spectrum capabilities without large forces or sophisticated training capabilities, like the U.S. CTCs. As a result, as we hope we demonstrate in this monograph, the United States can learn much from the experiences of these other nations.

Bibliography

Abrial, Stéphane, "A Highly Professional Air Force," *NATO's Nations and Partners for Peace*, Vol. 52, No. 2, 2007.

Acosta, Marcus P., "The Kargil Conflict: Waging War in the Himalayas," *Small Wars & Insurgencies*, Vol. 18, No. 3, September 2007.

"Air Force, India," *Jane's Sentinel Security Assessment—South Asia*, February 13, 2008.

Air Force Special Operations Command, "AFSOC Combat Aviation Advisor Mission Qualification Course Formal Training Pipeline," undated.

Allen, Kenneth, "The PLA Air Force: 2006–2010," paper presented at the "CAPS-RAND-CEIP International Conference on PLA Affairs," Taipei, 2005.

Allen, Kenneth, and Mary Ann Kivlehan-Wise, "Implementing PLA Second Artillery Doctrinal Reforms," in James Mulvenon and David Finkelstein, eds., *China's Revolution in Doctrinal Affairs: Emerging Trends in the Operational Art of the People's Liberation Army*, Alexandria, Va.: CNA Corporation, 2005.

Amnesty International, "Côte d'Ivoire: Clashes Between Peacekeeping Forces and Civilians: Lessons for the Future," October 5, 2005.

———, "Côte d'Ivoire: Threats Hang Heavy over the Future," October 13, 2005.

Bagayoko, Niagale, and Anne Kovacs, *La Gestion Interministérielle des Sorties de Conflits*, Centre d'Études en Sciences Sociales de la Défense, 2007.

Ben-David, Alon, "All Quiet on the Eastern Front, so Israel Will Revise IDF Organization and Doctrine," *Jane's International Defence Review*, March 1, 2004.

———, "Debriefing Teams Brand IDF Doctrine 'Completely Wrong,'" *Jane's Defence Weekly*, January 3, 2007.

———, "IDF Shifts Focus to Ground Forces," *Jane's Defence Weekly*, January 10, 2007.

———, "IDF Resumes Training in the Golan Heights," *Jane's Defence Weekly*, March 2, 2007.

———, "IDF Ponders Navy or Coast Guard Role," *Jane's Defence Weekly*, June 13, 2007.

Bennett, John T., "Mullen: U.S. Military Needs Larger Slice of GNP to Modernize," *Defense News*, November 28, 2007.

Berrette, Valerie, and Benoit Saint Vincent, *Implantation Locale des Régiments: L'Expérience des Régiments Anciennement Professionnalisés*, Centre d'Études en Sciences Sociales de la Défense, September 2003.

Bhohsle, Rahul K., "India's National Aspirations and Military Capabilities: A Prognostic Survey," in Vijay Oberoi, ed., *Army 2020: Shape, Size, Structure and General Doctrine for Emerging Challenges*, New Delhi: Knowledge World, 2005.

Blasko, Dennis J., *The Chinese Army Today: Tradition and Transformation for the 21st Century*, London: Routledge, 2006.

———, "PLA Ground Force Modernization and Mission Diversification: Underway in All Military Regions," in Roy Kamphausen and Andrew Scobell, eds., *Right-Sizing the People's Liberation Army: Exploring the Contours of China's Military*, Carlisle, Pa.: U.S. Army War College, 2007.

———, "PLA Conscript and Noncommissioned Officer Individual Training," in Roy Kamphausen, Andrew Scobell, and Travis Tanner, eds., *The "People" in the PLA: Recruitment, Training, and Education in China's Military*, Carlisle, Pa.: Strategic Studies Institute, 2008.

Boëne, Bernard, Thierry Nogues, and Saïd Haddad, "À Missions Nouvelles des Armées, Formation Nouvelle des Officiers des Armes? Enquête sur l'Adaptation de la Formation Initiale des Officiers des Armes aux Missions d'Après-Guerre Froide et à la Professionnalisation," Centre d'Études en Sciences Sociales de la Défense, 2001.

Braddock, Joe, and Ralph Chatham, *Report of the Defense Science Board Task Force on Training Superiority and Training Surprise*, Washington, D.C.: Office of the Under Secretary of Defense for Acquisition, Technology & Logistics, 2001.

Brannen, Kate, "Ft. Hood Commander Concerned Army Is Losing Full-Spectrum Capabilities," InsideDefense.com, April 9, 2009.

Brilliant, Joshua, "Analysis: Reservists Demand War Probe," UPI.com, August 21, 2006. As of March 12, 2008:
http://www.upi.com/International_Intelligence/Analysis/2006/08/21/analysis_reservists_demand_war_probe/9815/

Britains–SmallWars.com, "UK Forces Deployed in Sierra Leone," Web page, 2008. As of February 2008:
http://www.britains-smallwars.com/Sierraleone/forces.html

British Army, home page, undated. As of March 2008:
http://www.army.mod.uk/

————, *Officer Career Development*, July 1, 2003.

British Army, Land Warfare Centre, Operational Training and Advisory Group, "Operational Training and Advisory Group Brief," briefing, 2008.

Brom, Shlomo, "Political and Military Objectives in a Limited War Against a Guerilla [sic] Organization," in Shlomo Brom and Meir Elran, eds., *The Second Lebanon War: Strategic Perspectives*, Tel Aviv: Institute for National Security Studies, 2007.

Brom, Shlomo, and Meir Elran, eds., *The Second Lebanon War: Strategic Perspectives*, Tel Aviv: Institute for National Security Studies, 2007.

Camy, Olivier, "Cours de Droit Constitutionnel Général," Web page, undated. As of February 6, 2008:
http://www.droitconstitutionnel.net/cours_dcVemepratique.htm

Casey, George, "CSA Sends—Transition Team Commanders (Unclassified)," June 17, 2008.

Catignani, Sergio, *Israeli Counter-Insurgency and the Intifadas: Dilemmas of a Conventional Army*, London: Routledge, 2008.

Challans, Tim, "Emerging Doctrine and the Ethics of Warfare," paper presented at the "Joint Services Conference on Professional Ethics," Fort Leavenworth, Kan., 2006.

Charpentier, Herve, email to Stephen Arata, November 7, 2007.

Chary, "SCIPIO V1: Future Training Tool for Major Unit CPs within the French Army," *Objectif Doctrine*, No. 22-02/2001, February 2001.

China.org.cn, "[China's National Defense in 2002:] Appendix II: Major Military Exchanges with Other Countries in 2001–2002," Web page, undated. As of March 7, 2008:
http://www.china.org.cn/e-white/20021209/AppendixII.htm

————, "[China's National Defense in 2004:] Appendix III: Major Military Exchanges with Other Countries (2003–2004)," Web page, undated. March 7, 2008:
http://www.china.org.cn/e-white/20041227/AppendixIII.htm

————, "[China's National Defense in 2004:] Appendix V: Joint Exercises with Foreign Armed Forces (2003–2004)," Web page, undated. As of March 7, 2008:
http://www.china.org.cn/e-white/20041227/AppendixV.htm

————, "[China's National Defense in 2006:] IV. The People's Liberation Army," Web page, undated. As of March 11, 2007:
http://www.china.org.cn/english/features/book/194482.htm

————, "[China's National Defense in 2006:] Appendix II: Major International Exchanges of the Chinese Military 2005–2006," Web page, undated. As of March 7, 2008:
http://www.china.org.cn/english/China/194339.htm

————, "[China's National Defense in 2006:] Appendix V: Participation in UN Peacekeeping Operations (up to Nov. 30, 2006)," Web page, undated. As of March 7, 2008:
http://www.china.org.cn/english/China/194350.htm

————, "Who's Who in China's Leadership," Web page, undated. As of January 10, 2008:
http://www.china.org.cn/english/features/leadership/86673.htm

Cohen, Eliot A., Michael J. Eisenstadt, and Andrew J. Bacevich, *Knives, Tanks and Missiles: Israel's Security Revolution*, Washington, D.C.: Washington Institute for Near East Policy, 1998.

Cohen, Eliot A., and John Gooch, *Military Misfortunes: The Anatomy of Failure in War,* New York: The Free Press, 1990.

Cohen, Stephen Philip, *India: Emerging Power*, Washington, D.C.: Brookings Institution Press, 2001.

Cohen, Stuart A., "The Israel Defense Forces (IDF): From a 'People's Army' to a 'Professional Military'—Causes and Implications," *Armed Forces and Society*, Vol. 21, No. 2, Winter 1995.

Cole, Bernard D., "Chinese Naval Exercises and Training, 2001–2005: Reporting and Analysis," paper presented at the "CAPS-RAND-CEIP International Conference on PLA Affairs," Taipei, 2005.

Cone, Robert W., "The Changing National Training Center," *Military Review*, May–June 2006.

Corbett, John, Edward O'Dowd, and David Chen, "Building the Fighting Strength: PLA Officer Accession, Education, Training, and Utilization," in Roy Kamphausen, Andrew Scobell, and Travis Tanner, eds., *The "People" in the PLA: Recruitment, Training, and Education in China's Military*, Carlisle, Pa.: Strategic Studies Institute, 2008.

Cordesman, Anthony, *The British Defeat in the South and the Uncertain Bush 'Strategy' in Iraq: 'Oil Spots,' 'Ink Blots,' 'White Space,' or Pointlessness?* Washington, D.C.: Center for Strategic and International Studies, 2007.

————, "The 'Gaza War': A Strategic Analysis," final review draft, Center for Strategic and International Studies, February 2, 2009.

Coutau-Bégarie, Hervé, "La Sauvegarde Maritime: Réflexions sur un Nouveau Concept," *Revue Maritime*, No. 463, September 17, 2007.

Coutsoukis, Photius, "Israel Training," Photius.com, Web page, undated. As of April 7, 2008:
http://www.photius.com/countries/israel/national_security/israel_national_security_training.html

van Creveld, Martin, *The Sword and the Olive: A Critical History of the Israeli Defense Force*, New York: Public Affairs, 2002.

Defence Academy of the United Kingdom, "Higher Command and Staff Course," Web page, 2009. As of March 21, 2009:
http://www.da.mod.uk/colleges/jscsc/Courses/HCSC

Denny, Ken, "Alaskans Train at Top Jungle Warfare School in India," *National Guard Bureau News*, June 21, 2004.

Directorate of Reserve Forces and Cadets, *Future Use of the UK's Reserve Forces*, February 7, 2005.

de Durand, Etienne, and Bastien Irondelle, *Stratégie Aérienne Comparée: France, États-Unis, Royaume-Uni*, Centre d'Études en Sciences Sociales de la Défense, 2006.

École Militaire de Spécialisation de l'Outre-Mer et de l'Étranger, home page, undated. As of February 6, 2008:
http://www.defense.gouv.fr/terre/layout/set/popup/content/view/full/38385

―――, "Command Briefing," Paris, March 3, 2008.

Embassy of India in Lao PDR, "Assistance to Lao PDR," Web page, 2007. As of February 20, 2008:
http://indemblao.nic.in/assistance_relations.htm

"English Summary of the Winograd Commission Final Report," NYTimes.com, January 30, 2008. As of January 30, 2008:
http://www.nytimes.com/2008/01/30/world/middleeast/31winograd-web.html?_r=2&oref=slogin&pagewanted=print&oref=sloginon

Feldman, Yotam, "Israeli Army Recuperates Radical Theory," *Ha'aretz* (Israel), October 26, 2007.

Forget, Michel, "Spécificité du Rôle et des Contraintes des Forces Aériennes," *Penser les Ailes Françaises*, No. 13, April 2007.

Fourteenth Lok Sabha, Standing Committee on Defence, *Demands for Grants (2007–2008)*, April 2007.

des Francs, Colas, "Centre de Préparation des Forces," briefing, August 2, 2007.

French Embassy, "2003–2008 Military Programme Bill of Law," undated. As of August 2007:
http://www.ambafrance-us.org/atoz/mindefa.pdf

Government of China, "中华人民共和国中央军事委员会 [Central Military Commission of the People's Republic of China]," Web page, March 15, 2008. As of January 6, 2009:
http://www.gov.cn/test/2008-03/15/content_921057.htm

Gregory, Shaun, "France and *Missions de Paix*," *RUSI Journal*, Vol. 145, No. 4, August 2000.

Harel, Amos, and Avi Issacharoff, *34 Days: Israel, Hezbollah, and the Lebanon War*, New York: Palgrave Macmillan, 2008.

Hart, M. Joël, *Avis Présenté au Nom de la Commission de la Défense Nationale et des Forces Armées sur le Projet de Loi de Finances pour 2007 (No. 3341)*, Vol. IV, *Défense: Préparation et Emploi des Forces, Forces Terrestres*, 2006.

Hendel, Yoaz, "The Reserves Comeback," *Strategic Assessment*, Vol. 10, No. 4, February 2008.

Hoffman, Bruce, "The Logic of Suicide Terrorism," *The Atlantic Monthly*, Vol. 291, No. 5, June 2003.

Hoffman, Frank G., "Hizbollah and Hybrid Wars: U.S. Should Take Hard Lesson From Lebanon," *Defense News*, August 14, 2006.

———, *Conflict in the 21st Century: The Rise of Hybrid Wars*, Arlington, Va.: Potomac Institute for Policy Studies, 2007.

———, "Neo-Classical Counterinsurgency?" *Parameters*, Summer 2007.

———, "Hybrid Warfare and Challenges," *Joint Force Quarterly*, No. 52, 1st Quarter 2009.

Houqin, Wang, and Zhang Xingye [王厚卿 and 张兴业], eds.,《战役学》 [*Campaign Studies*], Beijing: National Defense University Press, 2000.

House of Commons, "Defence (Options for Change)," House of Commons Debate, July 25, 1990, *Parliamentary Debates*, Commons, 5th ser., Vol. 177, cols. 446–486.

———, "Defence Debate," July 14, 1994, *Parliamentary Debates*, Commons, 5th ser., Vol. 246, col. 1169.

———, "Statements in the House of Commons," November 12, 2007, *Parliamentary Debates*, Commons, 5th ser., Vol. 467, col. 446.

———, "Statements in the House of Commons," January 10, 2008, *Parliamentary Debates*, Commons, 5th ser., Vol. 470, col. 601.

House of Commons, Committee of Public Accounts, *Recruitment and Retention in the Armed Forces: Thirty-Fourth Report of Session 2006–07*, London: Her Majesty's Stationery Office, 2007.

House of Commons, Defence Committee, *Ninth Report: Statement on the Defence Estimates 1993*, London: Her Majesty's Stationery Office, 1993.

————, *Implementation of Lessons Learned from Operation Granby, 1993–94*, London: Her Majesty's Stationery Office, 1994.

————, *Eighth Report: The Strategic Defence Review*, London: Her Majesty's Stationery Office, 1998.

————, *Third Report: Lessons of Iraq*, London: Her Majesty's Stationery Office, 2004.

————, *Iraq: An Initial Assessment of Post-Conflict Operations*, London: Her Majesty's Stationery Office, 2005.

————, *UK Operations in Iraq: Thirteenth Report of Session 2005–06*, London: Her Majesty's Stationery Office, 2006.

————, *First Report: UK Land Operations in Iraq 2007*, London: Her Majesty's Stationery Office, 2007.

————, *Thirteenth Report: UK Operations in Afghanistan*, London: Her Majesty's Stationery Office, 2007.

————, *UK Land Operations in Iraq 2007: First Report of Session, 2007–2008*, London: Her Majesty's Stationery Office, 2007.

House of Commons, Select Committee on Defence, "Written Answers for 10 October 2007," October 10, 2007, *Parliamentary Debates*, Commons, 5th ser., Vol. 464, col. 636W.

HQ Air Command, "Deployed Operations," briefing, January 2008.

Inbar, Efraim, "How Israel Bungled the Second Lebanon War," *Middle East Quarterly*, July 1, 2007.

"India Tackles Rising Army Personnel Crisis," *Jane's Defence Weekly*, January 3, 2007.

Indian Army, *Doctrine for Sub Conventional Operations*, December 2006.

Indian Army, Army Training Command, *Indian Army Doctrine*, October 2004.

Indian Military Training Team in Bhutan, home page, undated. As of February 20, 2008:
http://indianarmy.nic.in/arimtrat.htm

Indian Navy, "Flag Officer Sea Training," Web site, undated. As of February 20, 2008:
http://indiannavy.nic.in/fost_main.htm

Integrated Defence Staff–India, "Welcome Message from Chief of Integrated Defence Staff," Web page, undated. As of February 19, 2008: http://ids.nic.in/welcome.html

Interagency Transformation, Education and Analysis, "ITEA Interagency Coordination Symposium Agenda," December 12–14, 2006.

International Crisis Group, "Côte d'Ivoire: The War Is Not Yet Over," Africa Report No. 72, November 28, 2003.

———, "Côte d'Ivoire: No Peace in Sight," Africa Report No. 82, July 12, 2004.

———, "Côte d'Ivoire: Le Pire Est Peut-Être à Venir," Africa Report No. 90, March 24, 2005.

International Institute for Strategic Studies, *The Military Balance 2008*, London: Routledge, 2008.

Inter-Services Institutions, "Defence Services Staff College (DSSC)," Web page, undated. As of February 19, 2008: http://armedforces.nic.in/interservice/isidssc.htm

Irondelle, Bastien, "Civil Military Relations and the End of Conscription in France," *Security Studies*, Vol. 12, No. 3, Spring 2003.

Israel Defense Forces, "Main Doctrine," Web page, undated. As of January 30, 2008: http://dover.idf.il/IDF/English/about/doctrine/main_doctrine.htm

"Israel Scales Back Gaza Operation," BBC News, March 10, 2008.

Israeli Ministry of Foreign Affairs, "Summary of IDF Operations Against Hizbullah in Lebanon," Web page, August 5, 2006. As of March 13, 2008: http://www.mfa.gov.il/MFA/Terrorism-+Obstacle+to+Peace/ Terrorism+from+Lebanon-+Hizbullah/Summary+of+IDF+operations+against+Hiz bullah+in+Lebanon+5-Aug-2006.htm

———, "Winograd Commission Submits Interim Report," Web page, April 30, 2007. As of February 18, 2008: http://www.mfa.gov.il/MFA/Government/Communiques/2007/Winograd+Inquir y+Commission+submits+Interim+Report+30-Apr-2007.htm?DisplayMode=print

Jane's World Armies, "Israel," Web page, date not available. As of March 13, 2008: http://www8.janes.com/Search/documentView.do?docId=/content1/ janesdata/binder/jwar/jwara174.htm@current&pageSelected=allJan es&keyword=Teffen%202012&backPath=http://search.janes.com/ Search&Prod_Name=JWAR&#toclink-j1131118924167154

Joint Forces Staff College, "Joint Interagency Multinational Planner's Course," JIMPC 07-1, November 13–17, 2006.

Joint Special Operations University, "SOF-Interagency Collaboration Course (Planning Draft)," July 21, 2006.

————, "Terrorism Response Senior Seminar Plan of Instruction," December 5–7, 2006.

Kamphausen, Roy, and Andrew Scobell, eds., *Right-Sizing the People's Liberation Army: Exploring the Contours of China's Military*, Carlisle, Pa.: U.S. Army War College, 2007.

Kamphausen, Roy, Andrew Scobell, and Travis Tanner, eds., *The "People" in the PLA: Recruitment, Training, and Education in China's Military*, Carlisle, Pa.: Strategic Studies Institute, 2008.

Katz, Yaakov, "Navy Officers Reprimanded over 'Hanit' Attack," *Jerusalem Post*, January 2, 2007.

————, "IDF Readying for Gaza Incursion—But Not Yet," *Jerusalem Post*, September 6, 2007.

————, "Soldiers Fought Bravely, in the Cause of a 'Mistaken Conception,'" *Jerusalem Post*, January 31, 2008.

"Kennedy's Fear of Iraq Civil War," BBC News, September 20, 2005.

Kergus, Jean, "La Prise en Compte des Spécificités de la Stabilisation dans les Exercices de Préparation et d'Évaluation," *Doctrine*, No. 12, August 2007.

Khalidi, Omar, *Khaki and the Ethnic Violence in India: Army, Police and Paramilitary Forces During Communal Riots*, Gurgaon, India: Three Essays Collective, 2003.

Kidon, Amir, "Chief Reservist Officer Speaks," Israel Defense Forces Web site, May 7, 2007. As of March 12, 2008: http://dover.idf.il/IDF/English/News/Up_Close/2007/May/0701.htm

Klein, Gary, *The Power of Intuition: How to Use Your Gut Feelings to Make Better Decisions at Work*, New York: Doubleday, 2003.

Kreps, Sarah E., "The 2006 Lebanon War: Lessons Learned," *Parameters*, Spring 2007.

Ladwig, Walter C., III, "A Hot Start for Cold Wars: The Indian Army's New Limited War Doctrine," *International Security*, Vol. 32, No. 3, Winter 2007–2008.

LeBot, Frank, "Licorne, or the Challenge to Reality," *Doctrine*, No. 9, June 2006.

Lefebvre, "Pourquoi les Armées Doivent-Elles se Doter d'un Centre d'Entrainement Tactique Permanent Interarmées et International en France?" *La Tribune du CID*, October 2006.

Libel, Tamir, "'Follow Me!': The IDF Officer School and the Changing Civil-Military Relations in Israel," briefing presented at "3rd International Conference," Military Psychology Center, Ramat Gan, Israel, February 20, 2009.

Ludra, Thakur Kuldip S., *Air Land Battle—The Indian Dichotomy: A Report Submitted to the Joint Chiefs of Staff Committee,* New Delhi: Institute for Strategic Research and Analysis, 2001.

MacFarland, Sean, Michael Shields, and Jeffrey Snow, "White Paper for CSA: The King and I—The Impending Crisis in Field Artillery's Ability to Provide Fire Support to Maneuver Commanders," undated [2008].

Matthews, Matt M., *We Were Caught Unprepared: The 2006 Hezbollah-Israeli War,* The Long War Series, Occasional Paper 26, Fort Leavenworth, Kan.: Combat Studies Institute Press, 2008.

Ministère de la Défense, "Les Objectifs Stratégiques de la France," Web page, undated. As of July 25, 2007:
http://www.defense.gouv.fr/defense/enjeux_defense/politique_de_defense/objectifs_strategiques/les_objectifs_strategiques_de_la_france

———, "Livre Blanc 1994–2003," Web page, undated. As of September 10, 2007:
http://www.defense.gouv.fr/defense/enjeux_defense/politique_de_defense/introduction/politique_de_defense/livre_blanc_1994_2003

———, PIA 00.200, *Doctrine Interarmées d'Emploi des Forces Armées en Opération,* 2003.

———, *Annuaire Statistique de la Défense—2006,* December 2006. As of August 22, 2007:
http://www.defense.gouv.fr/sga/decouverte/statistiques/annuaire_statistique_de_la_defense/annuaire_statistique_de_la_defense_2006

Ministère de la Défense, Armée de l'Air, "Missions," Web page, undated. As of February 6, 2008:
http://www.defense.gouv.fr/air/au_service_de_la_defense/missions/les_missions_de_l_armee_de_l_air

Ministère de la Défense, Armée de Terre, TTA 901, *Forces Terrestres en Opérations,* 1999.

———, *Doctrine d'Emploi des Forces Terrestres en Stabilisation,* November 2006.

———, "Towards a Professional Army 2008," briefing, 2008.

Ministère de la Défense, Armée de Terre, Centre de Doctrine d'Emploi des Forces, *Doctrine du Processes, Reception, Stationnement, Mouvement, Integration,* 2006.

———, FT-01, *Gagner la Bataille: Conduire à la Paix,* January 2007.

Ministère de la Défense, Armée de Terre, Centre de Doctrine d'Emploi des Forces, Division Recherche et Retour d'Expérience, *Cahier du Retex: Enseignements de l'Opération Licorne*, July 2004.

Ministère de la Défense, État-Major des Armées, "Carte des OPEX [Opérations Extérieure]," Web page, undated. As of August 2007:
http://www.defense.gouv.fr/ema/operations_exterieures/carte_des_opex

—————, "Exercises Interarmées," Web page, undated. As of December 12, 2008:
http://www.defense.gouv.fr/ema/forces_interarmees/exercices_interarmees

—————, PIA 00.102, *Concept du Niveau Opératif,* July 2004.

—————, "Côte d'Ivoire: A l'Ouest, du Nouveau," Web page, July 25, 2007. As of October 26, 2009:
http://www.defense.gouv.fr/ema/operations_exterieures/cote_d_ivoire/breves/25_05_07_cote_d_ivoire_a_l_ouest_du_nouveau

—————, "Côte d'Ivoire: Opération 'Wagram,'" Web page, July 23, 2007. As of October 26, 2009:
http://www.defense.gouv.fr/ema/operations_exterieures/cote_d_ivoire/breves/23_07_07_cote_d_ivoire_operation_wagram

Ministère de la Défense, État-Major des Armées, Division d'Emploi, *Concept d'Emploi des Forces*, revised edition, October 8, 1997.

Ministère de la Défense, Gendarmerie, "Missions de Maintien de la Paix," undated. As of August 2007:
http://www.defense.gouv.fr/gendarmerie/enjeux_defense/les_engagements/operations_exterieures/missions_de_maintien_de_la_paix__1/missions_de_maintien_de_la_paix

Ministère de la Défense, Marine Nationale, "Dossier d'Information Marine 2006: Les Missions de la Marine—La Sauvegarde Maritime," Web page, undated. As of September 6, 2007:
http://www.defense.gouv.fr/marine/decouverte/organisation/dossier_d_information_marine_dim/dossier_d_information_marine_2006

—————, "Le Commandement Opérationnel: La Conduite des Forces," Web page, undated. As of September 2007:
http://www.defense.gouv.fr/marine/base/fichiers_presentation/le_commandement_operationnel_la_conduite_des_forces

—————, "Mise en Conditon [sic] Opérationnelle dans le Pacifique," Web page, undated. As of September 2007:
http://www.defense.gouv.fr/marine/decouverte/activites/exercices/mise_en_conditon_operationnelle_dans_le_pacifique

—————, "Trident d'Or 05," Web page, undated. As of September 2007:
http://www.defense.gouv.fr/marine/decouverte/activites/exercices/trident_d_or_05

Ministry of Defence, Government of India, *Annual Report, 2006–2007*, undated.

Ministry of National Defense, "[China's National Defense in 2006:] Appendix IV: Joint Exercises with Foreign Armed Forces (2005–2006)," Web page, January 15, 2007. As of September 9, 2009:
http://eng.mod.gov.cn/Database/WhitePapers/2007-01/15/content_4004332.htm

———, "[China's National Defense in 2008:] Appendix I: Major International Exchanges of the Chinese Military (2007–2008)," Web page, July 31, 2009. As of October 19, 2009:
http://eng.mod.gov.cn/Database/WhitePapers/2009-07/31/content_4017029.htm

———, "[China's National Defense in 2008]: Appendix II: Joint Exercises and Training with Foreign Armed Forces (2007–2008)," Web page, July 31, 2009. As of October 19, 2009:
http://eng.mod.gov.cn/Database/WhitePapers/2009-07/31/content_4017039.htm

———, "[China's National Defense in 2008]: Appendix III: China's Participation in UN Peacekeeping Operations," Web page, July 31, 2009. As of October 19, 2009:
http://eng.mod.gov.cn/Database/WhitePapers/2009-07/31/content_4017040.htm

Morinière, Frédéric, Guillame Danes, and Laurent Tard, "La Stabilisation: Encore Plus d'Exigence pour les Simulations," *Doctrine*, No. 12, August 2007.

Mulvenon, James, and David Finkelstein, eds., *China's Revolution in Doctrinal Affairs: Emerging Trends in the Operational Art of the People's Liberation Army*, Alexandria, Va.: CNA Corporation, 2005.

Murray, Williamson, and Richard Hart Sinnreich, eds., *The Past as Prologue: The Importance of History to the Military Profession*, New York: Cambridge University Press, 2006.

Naskari, N. S., "Doctrinal Changes and Imperatives of Force Restructuring," in Vijay Oberoi, ed., *Army 2020: Shape, Size, Structure and General Doctrine for Emerging Challenges*, New Delhi: Knowledge World, 2005.

National Air and Space Intelligence Center, *China: Connecting the Dots—Strategic Challenges Posed by a Re-Emergent Power*, Wright-Patterson AFB, Ohio: U.S. Air Force, 2007.

National Bureau of Statistics, *Statistical Yearbook of China 2006*, Beijing: China Statistics Press, 2006.

National Defence College, home page, undated. As of February 19, 2008:
http://ndc.nic.in/

Naveh, Shimon, *In Pursuit of Military Excellence: The Evolution of Operational Theory*, London: Frank Cass, 1997.

"NAVFCO—The French Navy's Arm for the Training of Friendly Foreign Navies," *Asian Defence Journal*, December 1999.

"Navy, India," *Jane's Sentinel Security Assessment—South Asia*, January 3, 2008.

Neveux, Bruno, "Command and Control for Operation Artemis," *Doctrine*, No. 5, December 2004.

"NORINCO Type 98/Type 99 MBT," *Jane's Armour and Artillery*, February 28, 2008.

Oberoi, Vijay, ed., *Army 2020: Shape, Size, Structure and General Doctrine for Emerging Challenges*, New Delhi: Knowledge World, 2005.

Office of Naval Intelligence, *China's Navy 2007*, Washington, D.C., 2007.

Pascallon, Pierre, ed., *Les Armées Françaises à l'Aube du XXIe Siècle*, Vol. III, *L'Armée de Terre*, Paris: Harmattan, 2003.

Pengelley, Rupert, "French Army Transforms to Meet the Challenge of Multirole Future," *Jane's International Defence Review*, June 2006.

———, "Reality Check: Learning the Art of War in an Age of Diverse Threats," *Jane's International Defence Review*, November 1, 2006.

Pfeffer, Anshel, "The Defense Establishment's Financial Brinkmanship," *Jerusalem Post*, August 28, 2006.

Ploch, Lauren, *Africa Command: U.S. Strategic Interests and the Role of the U.S. Military in Africa*, Washington, D.C.: Congressional Research Service, Report RL34003, 2007.

Porteret, Vincent, Emmanuelle Prevot, and Katia Sorin, *Armée de Terre et Armée de l'Air en Opérations: L'Adaptation des Militaires aux Missions*, Centre d'Études en Sciences Sociales de la Défense, 2006.

Pung, Hans, Laurence Smallman, Tom Ling, Michael Hallsworth, and Samir Puri, *Remuneration and Its Motivation of Service Personnel: Focus Group Investigation and Analysis*, Santa Monica, Calif.: RAND Corporation, DB-549-MOD, 2007. As of August 13, 2009:
http://www.rand.org/pubs/documented_briefings/DB549/

Qadir, Shaukat, "An Analysis of the Kargil Conflict 1999," *RUSI Journal*, April 2002.

Raoult, M. Eric, "Rapport Fait au Nom de la Commission des Affaires Étrangères sur la Proposition de Résolution No. 1968, Tendant à la Création d'une Commission d'Enquête sur les Conditions dans Lesquelles le Gouvernement Est Intervenu dans la Crise de Côte d'Ivoire Depuis le 19 Septembre 2002," Assemblée Nationale, Report No. 2032, January 18, 2005.

Rapaport, Amir, "The Unification Struck at Dawn," *Ma-ariv* (Israel), April 10, 2007, in BBC Monitoring, April 12, 2007.

Rayment, Sean, "UK's Last 1,000 Soldiers Rushed Out to Balkans," *Daily Telegraph* (London), February 17, 2008.

Ricks, Thomas E., *Fiasco: The American Military Adventure in Iraq*, New York: Penguin Press, 2006.

Riley, Jonathon, "The U.K. in Sierra Leone: A Post-Conflict Operation Success?" The Heritage Foundation, Heritage Lecture No. 958, August 10, 2006.

Romm, Giora, "A Test of Rival Strategies: Two Ships Passing in the Night," in Shlomo Brom and Meir Elran, eds., *The Second Lebanon War: Strategic Perspectives*, Tel Aviv: Institute for National Security Studies, 2007.

Rouby, Giles, "The Joint Dimension of Operations Command," *Doctrine*, No. 5, December 2004.

Royal Air Force, home page, undated. As of March 2008:
http://www.raf.mod.uk/

Royal Military Academy Sandhurst, "Overseas Cadets," Web page, undated. As of December 14, 2007:
http://www.sandhurst.mod.uk/courses/overseas.htm

Royal Navy, "Flag Officer Sea Training," Web page, undated. As of March 2008:
http://www.royalnavy.mod.uk/server/show/nav.3682

———, home page, undated. As of March 2008:
http://www.royalnavy.mod.uk

Sahgal, Arun, "National Military Aspirations and Military Capabilities: An Approach," in Vijay Oberoi, ed., *Army 2020: Shape, Size, Structure and General Doctrine for Emerging Challenges*, New Delhi: Knowledge World, 2005.

Schank, John F., Harry J. Thie, Clifford M. Graf II, Joseph Beel, and Jerry M. Sollinger, *Finding the Right Balance: Simulator and Live Training for Navy Units*, Santa Monica, Calif.: RAND Corporation, MR-1441-NAVY, 2002. As of August 13, 2009:
http://www.rand.org/pubs/monograph_reports/MR1441/

Sénat Français, "Projet de Loi, Adopté le 15 Janvier 2003, No. 49, Sénat, Session Ordinaire de 2002–2003, Projet de Loi Relatif à la Programmation Militaire pour les Années 2003 à 2008," January 2003. As of August 16, 2007:
http://www.senat.fr/leg/tas02-049.html

———, "Projet de Loi de Finances pour 2007: Défense—Préparation et Équipement des Forces: Forces Aériennes," 2006. As of August 16, 2008:
http://www.senat.fr/rap/a06-081-6/a06-081-613.html

Shrivastava, V. K., "Indian Air Force in the Years Ahead: An Army View," *Strategic Analysis*, Vol. 25, No. 8, November 2001.

Siboni, Gabriel, "The Military Campaign in Lebanon," in Shlomo Brom and Meir Elran, eds., *The Second Lebanon War: Strategic Perspectives*, Tel Aviv: Institute for National Security Studies, 2007.

————, "High Trajectory Weapons and Guerilla [sic] Warfare: Adjusting Fundamental Security Concepts," *Strategic Assessment*, Vol. 10, No. 4, February 2008.

Spiegel, Peter, "Army Is Training Advisors for Iraq," *Los Angeles Times*, October 25, 2006.

Spirtas, Michael, Jennifer D. P. Moroney, Harry J. Thie, Joe Hogler, and Thomas Durrell-Young, *Department of Defense Training for Operations with Interagency, Multinational, and Coalition Partners*, Santa Monica, Calf.: RAND Corporation, MG-707-OSD, 2008. As of August 13, 2009:
http://www.rand.org/pubs/monographs/MG707/

Storey, Ian, "Thai Massage for China's Military Muscle," *Asia Times Online*, July 11, 2008. As of December 18, 2008:
http://www.atimes.com/atimes/China/JG11Ad01.html

Tachon, Nicolas, "Educating for Military Operations in Urban Terrain," briefing to RAND Corporation staff, Tours, France, March 4, 2008.

Tessier, M. Guy, "Rapport Fait au Nom de la Commission de la Défense Nationale et des Forces Armées sur le Projet de Loi (No. 187) Relatif à la Programmation Militaire pour les Années 2003 à 2008," No. 383, November 25, 2002.

————, "Rapport d'Information Dépose en Application de l'Article 145 du Règlement par la Commission de la Défense Nationale et des Forces Armées en Conclusion des Travaux d'une Mission d'Information Constituée le 29 Mars 2005, sur le Contrôle de l'Exécution des Crédits de la Défense pour l'Exercice 2005," No. 2985, March 29, 2006.

Teule, Jean-Pierre, "Le CPCO au Cœur de Nos Opérations," *Revue Défense Nationale*, May 2007.

Thieblemont, André, Christophe Pajon, and Yves Racaud, *Le Métier de Sous-Officier dans l'Armée de Terre Aujourd'hui*, Centre d'Études en Sciences Sociales de la Défense, May 2004.

Thirteenth Lok Sabha, Standing Committee on Defence, *Manpower Planning and Management Policy in Defence*, August 24, 2001.

Tillson, John C. F., Waldo D. Freeman, William R. Burns, John E. Michel, Jack A. LeCuyer, Robert H. Scales, and D. Robert Worley, *Learning to Adapt to Asymmetric Threats*, Alexandria, Va.: Institute for Defense Analyses, D-3114, 2005.

"Tironut," Wikipedia, date not available. As of April 11, 2008:
http://en.wikipedia.org/wiki/Tironut

UK Defence Analytical Services Agency, *UK Armed Forces Quarterly Manning Report*, TSP 4, London, 2008.

UK Foreign & Commonwealth Office, *Active Diplomacy for a Changing World: The UK's International Priorities*, London: Her Majesty's Stationery Office, 2006.

UK Ministry of Defence, "Operations in Afghanistan: British Forces," Web page, undated. As of March 2008:
http://www.mod.uk/DefenceInternet/FactSheets/OperationsFactsheets/OperationsInAfghanistanBritishForces.htm

———, "Operations in Iraq: Facts and Figures," Web page, undated. As of September 14, 2009:
http://www.mod.uk/DefenceInternet/FactSheets/OperationsFactsheets/OperationsInIraqFactsandFigures.htm

———, *The United Kingdom Defence Programme: The Way Forward*, Cmnd. 8288, London: Her Majesty's Stationery Office, 1981.

———, *The Falklands Campaign: The Lessons*, Cmnd. 8758, London: Her Majesty's Stationery Office, 1982.

———, *Statement of Defence Estimates: Defending Our Future*, Cmnd. 2270, London: Her Majesty's Stationery Office, 1993.

———, *The Strategic Defence Review*, Cmnd. 3999, London: Her Majesty's Stationery Office, July 1998.

———, *The Strategic Defence Review: Supporting Essays*, Cmnd. 3999, London: Her Majesty's Stationery Office, 1998.

———, *The Strategic Defence Review: A New Chapter*, Vol. 1, Cmnd. 5566, London: Her Majesty's Stationery Office, 2002.

———, *Delivering Security in a Changing World*, Vol. 1, Cmnd. 6041-I, London: Her Majesty's Stationery Office, December 2003.

———, *Delivering Security in a Changing World: Supporting Essays*, Vol. 2, Cmnd. 6041-II, London: Her Majesty's Stationery Office, December 2003.

———, Joint Warfare Publication 3-00, *Joint Operations Execution*, 2nd ed., London: Her Majesty's Stationery Office, March 2004.

———, *Delivering Security in a Changing World: Future Capabilities*, Cmnd. 6269, London: Her Majesty's Stationery Office, July 2004.

———, *MOD Annual Report and Accounts 2006–7*, London: Her Majesty's Stationery Office, July 2007.

———, *UK Defence Statistics 2007*, London: Defence Analytical Services Agency, September 2007.

———, "Exclusive: Back to the Future: Army Training Is Ahead of the Game in Canada," Web page, October 29, 2007. As of December 10, 2007:
http://www.mod.uk/DefenceInternet/DefenceNews/TrainingAndAdventure/ExclusiveBackToTheFutureArmyTrainingIsAheadOfTheGameInCanadaaudio.htm

————, *The Military Balance*, Vol. 108, No. 1, February 2008.

————, "New Measures to Reward and Retain Forces Personnel," Web page, March 19, 2008. As of March 2008:
http://www.mod.uk/DefenceInternet/DefenceNews/DefencePolicyAndBusiness/NewMeasuresToRewardAndRetainForcesPersonnel.htm

UK Ministry of Defence, Development, Concepts and Doctrine Centre, JDP 4-00, *Logistics for Joint Operations*, 3rd ed., April 2007.

UK Ministry of Defence, HQ Land Forces, "Force Preparation and Generation," briefing, 2008.

UK National Audit Office, *Ministry of Defence: Assessing and Reporting Military Readiness*, London: Her Majesty's Stationery Office, 2005.

————, *Ministry of Defence: Recruitment and Retention in the Armed Forces*, London: Her Majesty's Stationery Office, 2006.

U.S. Department of the Army, "Military Transition Team OIF-OEF Training Model," Web page, undated. As of February 26, 2007:
http://www.riley.army.mil/view/document.asp?ID=727-2007-02-14-53396-3

————, FM 25-100, *Training the Force*, Washington, D.C.: Headquarters, Department of the Army, 1998.

————, Army Regulation 350-50, *The Army Combat Training Center Program*, Washington, D.C.: Headquarters, Department of the Army, January 24, 2003.

————, FM 1, *The Army*, Washington, D.C.: Headquarters, Department of the Army, 2005.

————, FM 22-103, *Leadership and Command at Senior Levels*, Washington, D.C.: Headquarters, Department of the Army, 2007.

————, FM 1-01, *Generating Force Support for Operations*, DRAG draft, April 2007.

————, *2008 Army Posture Statement: A Campaign Quality Army with Joint and Expeditionary Capabilities*, Washington, D.C.: Headquarters, Department of the Army, 2008.

————, FM 3-0, *Operations*, Washington, D.C.: Headquarters, Department of the Army, 2008.

————, "Battle Command Training Program," briefing, May 12, 2008. As of April 2, 2009:
https://www.us.army.mil/suite/doc/12921683

————, "TMAAG Concept Paper," May 12, 2008.

U.S. Department of the Army and U.S. Marine Corps, FM 3-24/MCWP 3-33.5, *Counterinsurgency Field Manual*, Chicago: The University of Chicago Press, 2007.

U.S. Department of Defense, *The National Defense Strategy of the United States of America*, Washington, D.C.: Department of Defense, 2005.

U.S. Department of Defense, Office of the Secretary of Defense, *Annual Report to Congress: Military Power of the People's Republic of China 2008*, undated [c. 2008].

————, *Annual Report to Congress: Military Power of the People's Republic of China,* 2009.

U.S. Department of Defense, Office of the Under Secretary of Defense for Personnel and Readiness, *Population Representation in the Military Services, FY 2004*, May 2006.

U.S. Government Accountability Office, *Military Training: Actions Needed to Enhance DoD's Program to Transform Joint Training*, GAO-05-548, Washington, D.C., 2005.

————, *Management Actions Needed to Enhance DOD's Investment in the Joint National Training Capability*, GAO-06-802, Washington, D.C., August 2006.

U.S. Joint Chiefs of Staff, CJCSM 3500.03A, *Joint Training Manual for the Armed Forces of the United States*, Washington, D.C.: Joint Staff, 2002.

————, CJCSI 3500.01C, *Joint Training Policy and Guidance for the Armed Forces*, Washington, D.C.: Joint Staff, 2004.

————, JP 3-0, *Joint Operations*, Washington, D.C.: Joint Staff, 2006.

U.S. Marine Corps, Marine Corps Training and Advisory Group, "U.S. Marine Corps Forces Command," briefing, October 2007.

————, "MCTAG Information Paper," May 12, 2008.

Vandergriff, Donald E., "Old Dogs for New Tricks," *Army*, November 2007/December 2007.

Vandergriff, Donald E., and George Reed, "Old Dogs and New Tricks: Setting the Tone for Adaptability," *Army*, August 2007.

Verstappen, Caroline, "Sociologie: Effet des Évolutions Démographiques et Sociales," in Pierre Pascallon, ed., *Les Armées Françaises à l'Aube du XXIe Siècle*, Vol. III, *L'Armée de Terre,* Paris: Harmattan, 2003.

Ward, Thomas E., II, "A JTF Training Dilemma: Component Rigor Versus Joint Realism," *Joint Forces Quarterly*, 3rd Quarter 2007.

Wegman, Yehuda, "The Struggle for Situation Awareness in the IDF," *Strategic Assessment*, Vol. 10, No. 4, February 2008.

Weitz, Richard, *The Reserve Policies of Nations: A Comparative Analysis*, Carlisle, Pa.: Strategic Studies Institute, 2007.

Wong, Leonard, *Stifled Innovation? Developing Tomorrow's Leaders Today*, Carlisle, Pa.: Strategic Studies Institute, 2002.

———, *Developing Adaptive Leaders: The Crucible Experience of Operation Iraqi Freedom*, Carlisle, Pa.: Strategic Studies Institute, 2004.

Wunderle, William, and Andre Briere, "U.S. Foreign Policy and Israel's Qualitative Military Edge," Washington Institute for Near East Policy, Policy Focus No. 80, January 2008.

Xiaoyang, Jiao, "Insight: Engineering Peace, Prosperity in Darfur," *China Daily* (online version), September 17, 2007. As of November 10, 2009: http://www.chinadaily.com.cn/cndy/2007-09/17/content_6110869.htm

Xinhua News Agency, "Chinese Troops Ready for UN Peace Mission," November 19, 2003. As of November 10, 2009: http://english.chinamil.com.cn/special/e-peace/txt/30.htm

Yost, David S., "France's Evolving Nuclear Strategy," *Survival*, Vol. 47, No. 3, Autumn 2005.

Youngs, Tim, and Mark Oakes, *Iraq: Desert Fox and Policy Developments*, International Affairs and Defence Section, House of Commons Library, Research Paper 99/13, February 10, 1999.

Yuilang, Zhang [张玉良], ed.,《战役学》[*Campaign Studies*], Beijing: National Defense University Press, 2006.